THE NATIONAL BUREAU OF STANDARDS

Functions and Activities

The functions of the National Bureau of Standards are set forth in the Act of Congress, March 3, 1901, as amended by Congress in Public Law 619, 1950. These include the development and maintenance of the national standards of measurement and the provision of means and methods for making measurements consistent with these standards; the determination of physical constants and properties of materials; the development of methods and instruments for testing materials, devices, and structures; advisory services to government agencies on scientific and technical problems; invention and development of devices to serve special needs of the Government; and the development of standard practices, codes, and specifications. The work includes basic and applied research, development, engineering, instrumentation, testing, evaluation, calibration services, and various consultation and information services. Research projects are also performed for other government agencies when the work relates to and supplements the basic program of the Bureau or when the Bureau's unique competence is required. The scope of activities is suggested by the listing of divisions and sections on the inside of the back cover.

Publications

The results of the Bureau's research are published either in the Bureau's own series of publications or in the journals of professional and scientific societies. The Bureau publishes three periodicals available from the Government Printing Office: The Journal of Research, published in four separate sections, presents complete scientific and technical papers; the Technical News Bulletin presents summary and preliminary reports on work in progress; and the Central Radio Propagation Laboratory Ionospheric Predictions provides data for determining the best frequencies to use for radio communications throughout the world. There are also five series of nonperiodical publications: Monographs, Applied Mathematics Series, Handbooks, Miscellaneous Publications, and Technical Notes.

A complete listing of the Bureau's publications can be found in National Bureau of Standards Circular 460, Publications of the National Bureau of Standards, 1901 to June 1947 ($1.25), and the Supplement to National Bureau of Standards Circular 460, July 1947 to June 1957 ($1.50), and Miscellaneous Publication 240, July 1957 to June 1960 (includes Titles of Papers Published in Outside Journals 1950 to 1959) ($2.25); available from the Superintendent of Documents, Government Printing Office, Washington D.C. 20402.

NATIONAL BUREAU OF STANDARDS

$\mathcal{T}$echnical $\mathcal{N}$ote 181

Issued August 20, 1963

COMPUTER PROGRAM FOR IONOSPHERIC
MAPPING BY NUMERICAL METHODS

Martha E. Hinds and William B. Jones

NBS Boulder Laboratories

Boulder, Colorado

NBS Technical Notes are designed to supplement the Bureau's regular publications program. They provide a means for making available scientific data that are of transient or limited interest. Technical Notes may be listed or referred to in the open literature.

TABLE OF CONTENTS

COMPUTER PROGRAM FOR IONOSPHERIC MAPPING

BY NUMERICAL METHODS

Martha E. Hinds and William B. Jones

A solution has recently been given to the problem of representing the complex variations of ionospheric characteristics on a world-wide scale, including their diurnal variation, by numerical analysis of ionospheric data as measured at a network of stations [Jones and Gallet, 1962a]. The present paper describes the IBM 7090 (FAP) program of the methods of numerical mapping referred to above. Included are detailed flow charts of the program logic, and all necessary information for applying the program. Thus it fills the gap between the publications giving the scientific bases for the methods of mapping and the practical problem of producing ionospheric maps. This program, applied to ionospheric characteristics foF2 and F2-M3000, forms the basis for the new series, Central Radio Propagation Laboratory Ionospheric Predictions.

1. INTRODUCTION

A solution has recently been given to the problem of representing the complex variations of ionospheric characteristics by numerical analysis of ionospheric data as measured at a network of stations [Jones and Gallet, 1962a].[1] The methods employed consist of well-defined mathematical operations which are readily adapted to high-speed automatic computing.[2] The principal output of these methods is a set of numerical coefficients defining a continuous function of latitude, longitude and time which represents the ionospheric characteristic on a world-wide scale, including its diurnal variation. The ionospheric representation thus obtained is called a numerical map.

---

[1] For a brief summary of the problem and its solution the reader can refer to [Jones and Gallet, 1960].

[2] The need for world-wide mapping methods based on numerical analysis and the use of high-speed computers has been felt for many years [C.C.I.R., 1959].

An integral part of the work in forming and testing the mapping methods referred to above was the parallel development and application of computer programs. This work evolved over a period of several years, beginning with an IBM 650 computer (in SOAP language), then with an IBM 704 (in SAP language) for its increased size and speed, and finally with the IBM 7090 (in FAP language). The present paper describes the 7090 program, referred to here as the Numerical Map Program.[1] The 7090 FAP listing of the Numerical Map Program is contained in [Hinds and Jones, 1962], which has been revised and condensed to form the present NBS Technical Note.

The numerical methods of mapping and computer programs referred to above, together with extensive applications made with the ionospheric characteristics foF2 and F2-M3000, form the basis for the new series,[2] Central Radio Propagation Laboratory Ionospheric Predictions, commencing in January 1963 [Ostrow, 1962; CRPL, 1963]. Among the advantages of this new series is that it is based on objective methods of mapping which are repeatable and well-defined in the mathematical sense, so that anyone starting with the same data will obtain an identical map. Although most of the mapping methods appear in published papers, there are many details of the mapping process which can be given only in a description of the computer program, for example, detailed flow charts of the program and all necessary information for applying the Numerical Map Program.

---

[1] Other 7090 computer programs (closely related to the one given here) have also been developed and extensively used. These include: (a) a Numerical Predictions Program for computing predicted numerical maps of ionospheric characteristics for future months (or days) and (b) a Contour Map Program for computing world-wide or polar contour maps in either universal time or local mean time.

[2] This series replaces the former CRPL Series D, "Basic Radio Propagation Predictions".

Flow Chart 1

NUMERICAL MAP PROGRAM

SUMMARY FLOW DIAGRAM OF MAIN PROGRAM STEPS

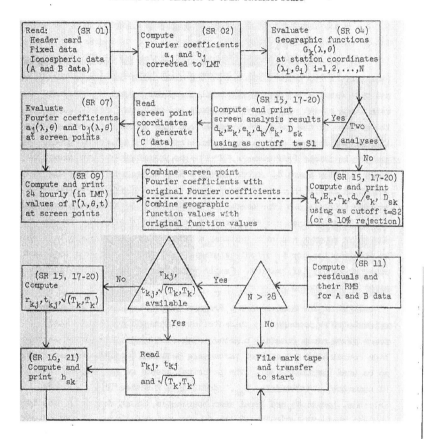

Thus, the purpose of the present paper is to fill the gap between the publications which give the scientific basis for the methods of mapping and the practical problem of producing ionospheric maps. Moreover, it is designed to complement the research papers already available by an overall treatment of the mapping processes, several of which are discussed here for the first time.

The backbone of the Numerical Map Program is an "executive program" (see section 2), which first enters the data using the routine 01 (SR 01) and then directs the operations until the final output is obtained. The main steps of the operations themselves are performed by 17 routines at the command of the executive program (see Flow Chart 1).[1] Each of these 17 routines forms a logical unit, which often is a relatively large problem in itself (for example, SR 17--the Gram-Schmidt Orthogonalization; see flow chart 8, second sheet). Even more, several routines are sometimes linked in a quasi-independent loop, with a loop executive program of their own (for instance, SR 20--Executive General Data Fitting). The 17 main routines are discussed in sections 3 through 9. For the executive program and each routine, a detailed program writeup is given to describe the input, output, storage, card formats, program logic, and other necessary information. Sample printouts of input and output are given in the appendices.

Input to the Numerical Map Program consists of the 24 hourly measurements of an ionospheric characteristic from all available stations for a given month (or day) together with such basic information as station coordinates and various parameters defining the mathematical functions to be used (section 2.3a). The principal output of the program is a set of numerical coefficients D_{sk}, defining a function $\Gamma(\lambda, \theta, t)$, of latitude, longitude, and local mean hour angle, which represents the ionospheric characteristic, including its diurnal variation, on a world-wide scale (section 2.3b).

---

[1]It will be noticed that in the numbering of the subroutines, numbers 3, 12, 13, and 14 are missing. These subroutines are not included since they are no longer used in the Numerical Map Program.

Although the basic theory for the methods of numerical mapping has been published [Jones and Gallet, 1962a], some essential procedures used in the Numerical Map Program have not been justified in scientific literature. In particular, the problem of the "stability" of the geographic representation in areas where no stations are available appears to be solved by a method of "screen points", which has been extensively applied and tested. While the method is described in section 2.2, further justification and illustrations will be given in a subsequent paper.

Section 9.1 describes a method for generating a special set of coefficients, h_{sk}, for intercomparing numerical maps from month to month or from day to day. Explanations of other new procedures and formulas are included in the text where they occur. It should also be mentioned that many of the routines in the present program were originally written for the 704 and are now retained with only slight modifications for purposes of economy. This is at no loss of program efficiency and accounts for certain differences of techniques appearing in the program.

This paper is intended for two types of users: (a) those who wish to apply the program as it stands, and (b) those who wish to use the program with modifications. For the former, a reading of only sections 2 and 3 will largely suffice. The remaining sections are included mainly for the latter. The terminology and notation used conform, whenever possible, with the reference papers [Jones and Gallet, 1962a and 1962b] so that most of the formulas and definitions are not repeated here. As a result, some knowledge of the above references will be needed for a thorough understanding of the program.

2. NUMERICAL MAP EXECUTIVE PROGRAM

2.1 Introduction

The primary function of the numerical map executive program is to direct the operation of the various subroutines. Section 2.2 discusses the concept and application of "screen points". Section 2.3 describes input, output, intermediate output on magnetic tape, storage requirements, and subroutines used. Section 2.3 also includes: (1) a summary list of sense switch settings, (2) a storage plan for the Numerical Map Program, (3) card formats for input and output, and (4) a flow chart of the numerical map executive program. Sample printouts of input and output are contained in appendices A and B, respectively. Before proceeding to these sections, however, it is desirable to review briefly the choice of time and geographic functions [Jones and Gallet, 1962a, chapter 3].

A number of options are available with regard to input, output, and types of analyses (section 2.3). The choice of the time and geographic functions to be used is considered here. A particular set of functions is specified by the following parameters entered on the header card (section 2.3a): (1) The total number of harmonics H used in the diurnal analysis. (2) The numbers k_0, k_1, and $k_2 = K$ (referred to in the program by the symbols PP1, PP2, and PP3 = P, respectively) which determine a particular set of geographic functions (table 1, section 5). Note that k_0 = PP1 must be greater than or equal zero. (3) In the geographic analyses, the residuals arrive at the noise level at a much lower degree for the high order harmonics than for the low order ones. Therefore, it is frequently necessary to truncate automatically the series of geographic functions for all harmonics of order higher than a certain value, e. g., H1. This can be done by entering H1 and a special cutoff, k = PH2, for the geographic series on the header card. (An example of values commonly used are H1 = 4 and PH2 = 4.) (4) Finally, a numerical code, FNCD, is used to identify the choice of time and geographic functions.

7

2.2 Screen Points

An important modification included in the present program is the
use of "screen points" for stabilizing the geographic representations
in areas where few or no measurements are available. As was pointed out
in chapter 5 of Jones and Gallet [1962a], the heavy grouping of stations
in some regions, such as Europe, and the absence of stations in other
regions tends to produce a sort of mathematical instability in the
representation function--that is, unrealistic behavior in areas where
no data are available. The best representation that can be expected for
such a region is a smooth continuation of the variations from surrounding
stations. The instability is alleviated by computing smooth continua-
tions at "screen points" chosen inside the gaps.[1] The "screen point
values" for the ionospheric characteristic are obtained from a deliber-
ately oversmoothed geographic representation of all the data. A second
analysis is then made using both the original data and the screen point
values (see flow chart 2).[2]

The oversmoothed representation in the first analysis is effected
by a radical truncation of the orthonormal series of geographic functions
[see section 4.2 of Jones and Gallet, 1962a]. By using a rejection
criterion considerably higher than 5 percent, all but the most significant terms are rejected. For example, the value $t = 4.0$ or 6.0 (for
the variable in Student's distribution) has been found to give a satis-
factory cutoff in the first analysis, whereas $t = 2.0$ is generally

---

[1]The set of screen points for the data of a given month is selected,
prior to using the program, in such a way as to fill all large gaps where
no stations are available and where consequently unacceptable instabil-
ities would be produced (section 2.3a, input (4)). In practice, the set
of screen points remains essentially unchanged from one month to the next.

[2]A slight variation in the method of defining the screen analysis
will be described in a later paper entitled "Representation of diurnal
and geographic variations of ionospheric data by numerical methods,
II. Control of instability." The screen analysis given there is made
by including only the main geographic variation.

8

Flow Chart 2

NUMERICAL MAP EXECUTIVE PROGRAM

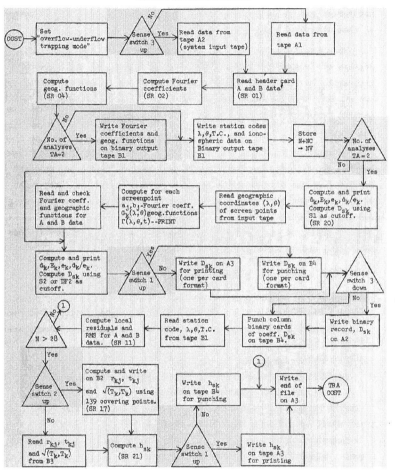

used in the second analysis. These values of t are denoted in the header card by S1 and S2, respectively. When it is not desired to use screen points, it is possible to make only one analysis. This option is also specified in the header card.

When the number of stations N is so small that the number of degrees of freedom $(DF = (N - k - 1)$, where $(k + 1)$ is the number of geographic functions) is less than or equal 28, it is assumed that only one analysis will be made, and the value of t is for a two tail, 10 percent rejection, depending on the number of degrees of freedom (see section 8.2d).

In addition to screen points described above, it has also been found helpful, in filling large gaps in the data, to use "predicted data" at certain stations for which data were not available for the given month or day being analyzed. Such "predictions" are made from correlations of measurements with a solar index. For purposes of identification in the present paper, we shall refer to measurements as A data, to predictions as B data, and to screen point values as C data.

2.3 Program Description

(a) Input (see sample printout, appendix A)

The following list gives the different types of input, ordered as they follow the program. Card formats and sample printouts are given for each type of card input.

(1) Header card (card format 1).

(2) Fixed data cards (card format 2)--see section 3.
A data cards precede B data cards.

(3) Ionospheric data cards (card format 3)--see section 3.
A data cards precede B data cards.

(4) Optional screen point coordinates on cards if a second analysis is to be made (card format 4).

(5) Optional input (on binary tape 3, channel B) r_{kj}, t_{kj}, and $\sqrt{(T_k, T_k)}$. In this case, these numbers can be computed in the first analysis and then entered as input in tape 3, channel B in each subsequent analysis by having sense switch 2 down.

The IBM 1401 is used to convert the binary program deck and the input data deck to magnetic tape.

If it is desired to use the Fortran monitor system, sense switch 3 must be up. In this case, tape 1, channel A is the system tape and the program and data are read from tape 2, channel A (the system input tape). There are two program options if the Fortran monitor system is not used: (1) if sense switch 3 is up, the self-loading program must be on tape 1, channel A and the input data must be on tape 2, channel A; (2) if sense switch 3 is down, the self-loading program followed by the input data must be on tape 1, channel A.

(b) Output. The following are regular and optional types of output:

(1) Tape 3, channel A, BCD output (see sample printout, appendix B).

First analysis:
d_k, E_k, e_k, d_k/e_k, $k = 0, 1, \ldots, K$ (for each Fourier coefficient).

Second analysis (optional):
Γ values at screen points and d_k, E_k, e_k, d_k/e_k. (The following are output from analysis 1 when there is no second analysis.) Coefficients D_{sk} (if sense switch 1 is down); residuals and their root mean square; intercomparison coefficients h_{sk} (if sense switch 1 is down).

(2) Tape 4, channel B, binary output to be punched into column binary cards by 1401 (see card formats 6 and 7). These cards contain the coefficients D_{sk} and are stored for permanent records and later use in predictions. First card is binary identification followed by the coefficients D_{sk}. These cards are punched in absolute locations for use in prediction programs.

11

(3) Optional binary output of coefficients D_{sk} is obtained on tape 2, channel A, when sense switch 3 is down. This output is generally used for immediate applications such as the computation of contour maps. (Note: if sense switch 3 is up, tape 2, channel A is used for input).

(4) Optional BCD tape 4, channel B will contain the coefficients D_{sk} and h_{sk} when sense switch 1 is up. These coefficients are then punched one per card for later use (see card format 5). Note: if sense switch 1 is down, these coefficients will be included in output 1 above.

(5) Optional binary output r_{kj}, t_{kj}, and $\sqrt{(T_k, T_k)}$ will be written on tape 2, channel B if sense switch 2 is up (see input 5 and also section 9).

(c) Intermediate output

(1) The geographic functions and Fourier coefficients for A and B data are written on tape 1, channel B if two analyses are used. This is done so that corresponding screen point results can be combined for the second analysis.

(2) Geographic latitudes and longitudes, time corrections, and ionospheric data (A and B data) are written on tape 1, channel B for subsequent use in the residual subroutine (SR 11).

(3) Tape 2, channel B is used to write the intermediate triangular matrices a_{kj}, b_{kj}, for use in SR 17 and SR 18 (see section 8). This tape is subsequently rewound and used for output if the option is chosen to compute the r_{kj}, t_{kj} (see output 5) rather than read them from input.

(d) Storage

At most instances during the operation of the program, the computer storage capacity of 32,000 + storage locations is used. See storage plan (h) for matrix limits and locations.

(e) Share subroutines used

Punch B, UA-SPHI, and UA-BDCI, along with those share subroutines incorporated in each of the seventeen subroutines.

(f) Remarks

Trapping mode is set for use in controlling underflow. End of file is written on tape 3, channel A after each set of data.

If consecutive months of data are to be run, push start at the program halt or return to system.

If at any time during the computation it is desired to see the off-line BCD output written on the on-line printer, put sense switch 6 down. The printing may again be suppressed by putting sense switch 6 up.

(g) Sense switch settings for Numerical Map Program

1 up   writes coefficients D_{sk} and h_{sk} (as above) on tape 4, channel B.

   down writes coefficients D_{sk} and h_{sk} in BCD one per card format on tape 3, channel A.

2 up   computes r_{kj}, t_{kj}, and $\sqrt{(T_k, T_k)}$ and writes output on tape 2, channel B. Computed values are used as input to SR 21.

   down assumes tape 3, channel B contains 3 binary records r_{kj}, t_{kj}, and $\sqrt{(T_k, T_k)}$ (see input (5)) for input to SR 21.

3 up   reads data (monitor system) from tape 2, channel A. Omits writing binary record of coefficients D_{sk} on tape 2, channel A.

   down reads program and data from tape 1, channel A. Writes binary record of coefficients D_{sk} on tape 2, channel A.

6 up   normal operating procedure.

   down BCD records are written on tape and the on-line printer.

Analysis 1. Max(N_A+N_B)=180 Max H=8 Max P=80	Analysis 2. MAX(N_A+N_B+N_C)=220 Max H=8 Max P=80

Fixed Data   5101(1st Loc)
   Station code   5101 ... +N
   Geog. long.   5281 ... +N
   Geog. lat.   5461 ... +N
   Time correction   5641 ... +N
 	
Ionospheric Data   5821
   Hour 0   5821 ... +N
   Hour 1   6001 ... +N
    ⋮   ⋮
   Hour 23   9961 ... +N
Maximum elements 5040 (=28 x 180)
	Fourier Coefficients 5101(1st Loc)
	a_0 5101 ... +(N+NC)
	b_1 5101+(N+NC) ... +(N+NC)
	a_1
	 ⋮ ⋮ ⋮
	b_H
	a_H 5101+2H(N+NC) ... +(N+NC)
	Maximum elements 3740 (=17 x 220)

Fourier Coefficients 11581(1st Loc) a_0 11581 ... +N b_1 (11581+N) ... +N a_1 ⋮ ⋮ ⋮ b_H a_H (11581+2HN) ... +N Maximum elements 3060 (=17 x 180)	Geographic Functions [M=5101+(2H+1)(N+NC)] (1st Loc) G_0 M ... +(N+NC) G_1 M+(N+NC) ... +(N+NC) ⋮ ⋮ G_P M+P(N+NC) ... +(N+NC) Maximum elements 17820(=81 x 220) Intermediate Storage 27228 Same order as Analysis 1

Geographic Functions
[M=11581+(2H+1)N] (1st Loc)
G_0 M ... +N
G_1 (M+N) ... +N
 ⋮   ⋮
G_P (M+PN) ... +N
Maximum elements 14580 (=81 x 180)
	Residual Subroutine (restored
	   from tape B1 as in Analysis 1)
	
	Fixed Data 5101(1st Loc)
	   Station code 5101 ... +N
	   Geog. long. 5281 ... +N
	   Geog. lat. 5461 ... +N
	   Time correction 5641 ... +N
	
	Ionospheric Data 5821
	   Hour 0 5821 ... +N
	   Hour 1 6001 ... +N
	    ⋮ ⋮
	   Hour 23 9961 ... +N
	Maximum elements 5040 (=28 x 180)

Intermediate Storage 27228(1st Loc) P(P+1)/2 Maximum elements 3240 Coefficients D_{sk}, 30468(1st Loc) Maximum elements 1377	Program Origin 101

14

Card Format 1

NUMERICAL MAP HEADER CARD

CARD COLUMN	DESCRIPTION		REMARKS
2	3		Header card
4-6	N	(xxx)	Number of stations (A and B data)
8-9	P	(xx)	P+1 = K+1 = Number of geographic functions
11	H1	(x)	Highest harmonic for which the geographic analysis is made with P+1 functions
13	H	(x)	Number of harmonics for diurnal analysis
15-17	FNCD	(xxx)	Function identification code
19-20	PP1	(xx)	k_0: Highest term for main latitudinal variation
22-23	PP2	(xx)	k_1: Highest term for first order longitudinal variation
25-26	PP3	(xx)	$k_2=K$: Highest term for second order longitudinal variation
28-29	PH2	(xx)	Highest term used in geographic analysis of harmonics $H1 < j \leq H$.
31-32	SF	($\pm$ n)	Scale factor
34-35	S1	$(xx)^1$	Student's t cutoff in first analysis
37-38	S2	$(xx)^1$	Student's t cutoff in second analysis
40	TA	(x)	Option of 1 or 2 analyses
42-44	NA	(xxx)	Number of stations with A data

[1] Program automatically places a decimal point between the first and second digits.

Card Format 1 (Cont.)

NUMERICAL MAP HEADER CARD

See previous page

CARD COLUMN	DESCRIPTION		REMARKS
46-48	NC	(xxx)	Number of screen points
50-53	S	(xxx.x)	Sunspot number
55-56	MN	(xx)	Order of months (or days) for use in prediction program
65-68	YYMM	(xxxx)	Year and month
69-72	DDLC	(xxxx)	Date, layer, and characteristic (see card format for ionospheric data)

16

Card Format 2

FIXED STATION DATA

CARD COLUMN	DESCRIPTION	REMARKS
1	1	Ionospheric data
2	0	Fixed station data
3-5	STC	Station code
6-9	000.0≤xxx.x<360.0	Geographic longitude (θ) in degrees east of Greenwich.
10-11	00	Constants
12	11 or 12 Punch	Sign of geographic latitude. 11 indicates southern latitude, 12 indicates northern latitude.
13-15	00.0≤xx.x≦90.0	Geographic latitude (λ) in degrees
16-19	000.0≤xxx.x< 360.0	Reference longitude (θ_R) for time zone See [Jones and Gallet, 1962a, section 3.1.]
20-21	00	Constants
22	11 or 12 Punch	Sign of time correction. 11 indicates negative correction, 12 indicates positive correction.
23-25	x.xx	Time correction (in hours). $T.C. = (\theta - \theta_R)/15.$
26-29	000.0≤xxx.x< 360.0	Geomagnetic longitude: see Smithsonian Physical Tables, Ninth Revised Edition, pp. 493-501.
30-31	00	Constants
32	11 or 12 Punch	Sign of geomagnetic latitude. 11 indicates south, 12 indicates north.
33-35	xx.x	Geomagnetic latitude: see Smithsonian Physical Tables, Ninth Revised Edition, pp. 493-501.

17

IONOSPHERIC DATA

CARD COLUMN	DESCRIPTION	REMARKS
1	1	Ionospheric data
2	1 or 2	1 refers to hours from 00:00 to 11:00 2 refers to hours from 12:00 to 23:00
3-5	STC	Station code
6-7	YY	Year
8-9	MM	Month
10-11	DD	Date: $01 \leq DD \leq 31$ indicates day of month $DD = 45$ indicates monthly median
12	L	Ionospheric layer: $L = 0$, F2-layer; $L = 1$, F1-layer; $L = 2$, E-layer; $L = 3$, E_s-layer
13	C	Ionospheric characteristic; $C = 0$, fo; $C = 1$, fx; $C = 3$, M3000; $C = 4$, h'; $C = 5$, h_p
14-73	Hourly Readings	Twelve readings of five positions each
		First four positions--numerical reading (R) Fifth position--qualifying symbol (S)

18

Card Format 4

SCREEN POINT INPUT

CARD COLUMN	DESCRIPTION	REMARKS
1, 7, ..., 73	s	Sign of geographic latitude
		(blank for plus and − for minus)
2-3, 8-9, ...74-75	λλ	Geographic latitude (degrees)
4-6, 10-12, ...76-78	θθθ	Geographic longitude (degrees east of Greenwich)

19

COEFFICIENTS D_{sk} AND h_{sk} DEFINING NUMERICAL
MAPS OF IONOSPHERIC CHARACTERISTICS

IDENTIFICATION					FUNCTIONS				SN	Coefficient D_{sk} or h_{sk}*			
YY	MM	DD	L	P	H	FNCD	PP1	PP2	T	xxx.x	ss	kk	±x.xxxxxxxE±yy

*For D_{sk} or h_{sk}, the number given by the first eight digits and sign is multiplied by 10 raised to the power given by the last two digits and sign.

CARD COLUMN	DESCRIPTION	REMARKS
4-5	YY	Year
6-7	MM	Month
11-12	DD	Date: See Ionospheric Data Card Format.
13	L	Ionospheric layer: See Ionospheric Data Card Format.
14	C	Ionospheric characteristic: See Ionospheric Data Card Format.
17-19	NNN	Number of stations used in the analysis
22-23	P	$P+1 = K+1 =$ Number of geographic functions
27	H	Number of harmonics for diurnal analysis
30-31	FNCD	Function identification code
34-35	PP1	k_0 = Highest term for main latitudinal variation
38-39	PP2	k_1 = Highest term for first order longitudinal variation
43	T	$T = 1$ for coefficients D_{sk} $T = 2$ for coefficients h_{sk}
47-51	xxx.x	Sunspot number
55-56 60-61	ss kk	Indices for coefficients D_{sk} or h_{sk} [Jones and Gallet, 1962a, chapter 7]
65-78	±x.xxxxxxxE±yy	Coefficient D_{sk} or h_{sk} in floating decimal form

Card Format 6

NUMERICAL MAP BINARY IDENTIFICATION

CARD COLUMN	DESCRIPTION	REMARKS
1	7 and 9 Punch	Column binary
2-3		Absolute binary location $(100)_{10}$
4-6	Check Sum	
7-9	YYMM	Year, month
10-12	DDLC	Date, layer, characteristic
13-15	N	Number of stations (A and B data)
16-18	P	P+1 = K+1 = Number of geographic functions
19-21	H	Number of harmonics for diurnal analysis
22-24	FNCD	Function identification code
25-27	PP1	k_0: Highest term for main latitudinal variation
28-30	PP2	k_1: Highest term for first order longitudinal variation
31-33	PP3	k_2 = K: Highest term for second order longitudinal variation

21

3. DATA READ SUBROUTINE 01

The data read subroutine (SR 01) is employed by the numerical map executive program for reading and storing the header card, fixed data, and ionospheric (A and B) data. Although most of the necessary information concerning input is given in section 2, certain additional information is necessary, in particular, the discussion of the <u>scale factor</u> and the list of <u>stops</u> resulting from improper input (section 3g). Other topics include: output, storage, calling sequence, subroutines used, general remarks, and program logic (flow chart 3).

(a) <u>Input</u> (see sample printout, appendix A)

Input card formats and sample printouts were discussed in section 2. One fixed data card and two ionospheric data cards must be read for each station (for A and B data). All fixed data cards precede the ionospheric data cards, and the latter must be paired by stations in the same order as the fixed data. In each pair, the type 1 card (col. 2) must precede the type 2 card (col. 2). Also, the stations with A data must precede those with B data.

For ionospheric data other than foF2, a special <u>scale factor</u> must be inserted in the header card. The ionospheric data are assumed to be given by four digit numbers (XXXX) (see card format 3). If no scale factor (blank) is punched on the header card, the program places a decimal point between the second and third digits (XX.XX). For other characteristics, however, (such as M3000) the position of the decimal point is specified by the scale factor. A scale factor of plus (or minus) n (n $\neq$ 0) moves the decimal point n places to the right (or left) from the position at the right of the fourth digit. "Minus zero" places the decimal point to the right of the fourth digit. For example, for M3000 the scale factor is -3 so that each ionospheric reading is taken as X.XXX.

(b) <u>Output</u> includes: station code, geographic longitude (θ) and latitude (λ), time correction, and ionospheric data.

23

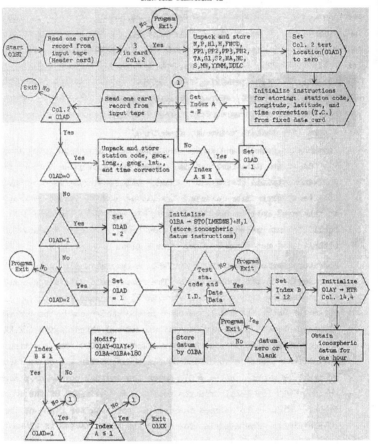

(c) Storage (see storage plan in section 2)

The input uses maximum storage of 5040 locations starting at $(5101)_{10}$. SR 01 itself takes 157 locations.

(d) Calling sequence

```
            C L A  (EXTIN)
            S T O  01XX
            T R A  01ST
   (EXTIN)  T R A  (NEXT)
   (NEXT)   (CONTINUE)
```

(e) Subroutines used

Modified share subroutine PE-CSMO, which includes: READT, FLOAT, FIXED, SCALE, and BCDPK.

(f) Remarks

(1) Index registers A, B, and C are used with no provision to restore them.

(2) In order to read data on tapes other than A1 or A2, a modification must be made in READT.

(g) Stops

SR 01 stops at 01AC for the following conditions:

(1) An ionospheric datum is zero or blank.

(2) Input cards not in proper order with respect to column 2. Proper order requires: (i) first card has 3 in column 2, (ii) the following N cards have 0 in column 2, and (iii) the following 2N cards have alternately 1 and 2 in column 2.

(3) Column 2 contains a number other than 3, 0, 1, 2.

(4) Identification of ionospheric data cards does not match identification on header card or ionospheric data cards (in pairs) are in different order from fixed data cards.

25

4. FOURIER ANALYSIS SUBROUTINE 02

SR 02 is used to compute Fourier coefficients for the diurnal analysis of the ionospheric data. The coefficients are then corrected to local mean time [Jones and Gallet, 1962a, section 3.1]. SR 02 is entered one time for each set of (A and B) ionospheric data (i.e., N times). This section gives a general description of the subroutine including the program logic (flow chart 4).

(a) Input

    (1) Hourly values y_1, y_2, ..., y_{24} of ionospheric data from one station.

    (2) Number H of harmonics.

    (3) Time correction T.C. (in hours).

(b) Output

Fourier coefficients a_0, a_j, and b_j, amplitudes c_j and phase angles ψ_j corrected to local mean time (LMT) for $j = 1, 2, ..., H$ (see [Jones and Gallet, 1962a], section 3.1). Note: the c_j and ψ_j are merely intermediate results and hence are not permanently stored.

(c) Storage

There are 4 (H + 1) storage locations required for output:
AO to AO + H, BO to BO + H, CO to CO + H, and PSIO to PSIO + H.
The routine itself (not including subroutines) takes 328 storage locations.

(d) Calling sequence

            C L A (EXTIN)

            S T O   02XX

   (EXTIN) T R A (NEXT)

   (NEXT) (CONTINUE)

(e) Subroutines used

Share subroutines: UASQR4, NAO331, UA‑ S + CL, and UA‑SPHI

Flow Chart 4

FOURIER ANALYSIS SUBROUTINE 02

(x = zone time hour angle; $x_i = (15°i-180°)$; i = 0,1,..., 23)

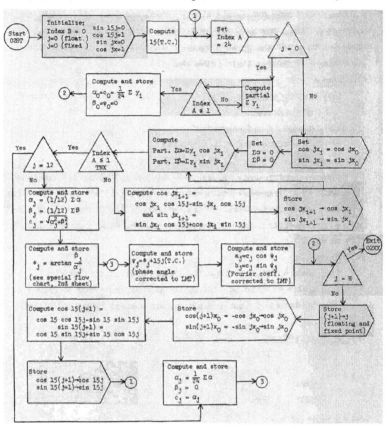

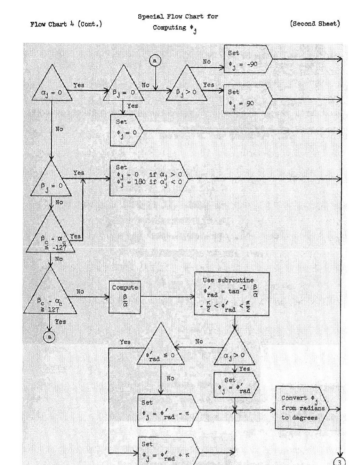

29

(f) Remarks

    (1) The AC and MQ overflow indicators should be off at entry to SR 02.

    (2) Index registers A, B, and C are used with no provision to restore them.

(g) Stops

A halt will occur at O2BM + 2 if an attempt is made to take the square root of a negative number.

5. GEOGRAPHIC FUNCTIONS SUBROUTINE 04

SR 04 is used to evaluate the geographic functions $G_k(\lambda, \theta)$ at station coordinates and at screen point coordinates (see table 1). These functional values are subsequently used in the geographic analyses. For this purpose, SR 04 is entered one time for each of the N stations and once for each screen point. SR 04 is again entered once for each of the N stations for computing residuals (section 7). A general description of the subroutine is given, including the program logic (flow chart 5).

(a) Input

  (1) Geographic coordinates (λ_i, θ_i) for one station (λ_i = latitude, θ_i = longitude).

  (2) Parameters PP1 (≥ 1), PP2, and PP3 = P defining the choice of geographic functions $G_k(\lambda, \theta)$. If no longitude terms are desired, we set PP1 = PP2 = PP3 = P. See table 1 (this section) and the discussion of the choice of functions in section 2.1.

  (3) Function identification code FNCD (see header card).

  (4) If 04ST entry is used, N, N-1, ..., 1 must be in index register A.

(b) Output

  Geographic function values $G_k(\lambda_i, \theta_i)$, k = 0, 1, ..., K.

(c) Storage

  (1) Entry through 04ST causes the $G_k(\lambda_i, \theta_i)$ to be stored in matrix form involving N(K+1) elements.

  (2) Entry through 04ST2 causes the $G_k(\lambda_i, \theta_i)$ to be stored in consecutive locations starting at FUNCT. This involves (K+1) elements.

  (3) SR 04 (not including subroutines) itself requires 286 storage locations.

31

TABLE 1

GEOGRAPHIC FUNCTIONS $G_k(\lambda,\theta)$

	k	$G_k(\lambda,\theta)$
Main Latitudinal Variation	0	1
	1	$\sin \lambda$
	2	$\sin^2 \lambda$
	...	...
	k_0	$\sin^{q_0} \lambda$
Mixed Latitudinal and Longitudinal Variation — First Order in Longitude	k_0+1	$\cos \lambda \cos \theta$
	k_0+2	$\cos \lambda \sin \theta$
	k_0+3	$\sin \lambda \cos \lambda \cos \theta$
	k_0+4	$\sin \lambda \cos \lambda \sin \theta$
	...	... ...
	...	... ...
	k_1-1	$\sin^{q_1}\lambda \cos \lambda \cos \theta$
	k_1	$\sin^{q_1}\lambda \cos \lambda \sin \theta$
Second Order in Longitude	k_1+1	$\cos^2\lambda \cos 2\theta$
	k_1+2	$\cos^2\lambda \sin 2\theta$
	k_1+3	$\sin \lambda \cos^2\lambda \cos 2\theta$
	k_1+4	$\sin \lambda \cos^2\lambda \sin 2\theta$
	...	... ...
	...	... ...
	$K-1$	$\sin^{q_2}\lambda \cos^2\lambda \cos 2\theta$
	K	$\sin^{q_2}\lambda \cos^2\lambda \sin 2\theta$

32

Flow Chart 5

GEOGRAPHIC FUNCTIONS SUBROUTINE O4

$$\begin{pmatrix} \text{LGK} & = \text{Loc. of } G_k(\lambda_i, \theta_i), \; i = 1,2, \ldots, N, \; k = 0,1, \ldots, P. \\ \text{FUNCT} & = \text{Loc. of } G_k(\lambda, \theta), \quad k = 0,1, \ldots, P. \end{pmatrix}$$

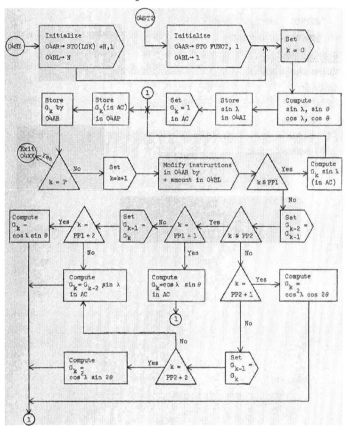

33

(d) Calling sequence

```
              C L A   EXTIN
              S T O   04XX
              T R A   04ST   (or 04ST2)
      EXTIN   T R A   (NEXT)
      (NEXT)  (CONTINUE)
```

(e) Subroutines used

Share subroutines: UA - S + CL and UA-SPHI

(f) Remarks

(1) The AC and MQ overflow indicators should be off at entry to SR 04.

(2) Index registers A, B, and C are used with no provision to restore them.

34

6. SYNTHESIS SUBROUTINES 05-10

6.1 Explanatory Remarks

Six subroutines are described in the present section. These subroutines are used to compute the value of the function $\Gamma(\lambda, \theta, t)$. One frequently needs to evaluate $\Gamma(\lambda, \theta, t)$, (a) at many locations for a fixed instant of time, or (b) at a fixed location for several different instants of time. In such cases, considerable savings in computer time can be gained by use of the following subroutines. The techniques employed are described in detail in section 3.1 of [Jones and Gallet, 1962b].

(a) SR 08 should be used when, for successive computations, Γ is evaluated for a fixed t and varying λ and θ. The entry for the first of such computations is 08ST, and for all following 08ST2 until t changes.

(b) SR 09 should be used for fixed λ and θ and varying t. The entry for the first of such computations is 09ST and for subsequent computations 09ST2 (until λ and θ change).

(c) SR 10 should be used for successive computations of Γ for fixed T and θ and for varying λ. SR 10 is entered initially at 10ST and then at 10ST2 until T and θ are changed.

The following three subroutines are used in connection with the above:

(d) SR 05 computes $\sin jt$ and $\cos jt$ for $j = 1, 2, \ldots, H$, given t, H, and TRA exit instruction in 05XX.

(e) SR 06 computes $D_k(t)$ for $k = 0, 1, \ldots, K$, given the matrix of coefficients D_{sk}, $\sin jt$ and $\cos jt$ $(j = 1, 2, \ldots, H)$, H, K, and TRA exit instruction in 06XX. For an explanation of $D_k(t)$, see section 3.1a of [Jones and Gallet, 1962b].

35

Flow Chart 6

SYNTHESIS SUBROUTINES 05-10

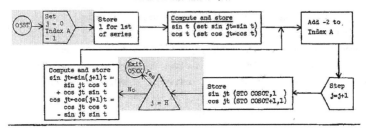

LDSK = Loc. of $D_k(t)$
LDM = Loc. of D_{sk} coefficients.

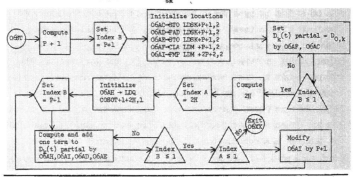

FUNCT = Loc. of $G_k(\lambda,\theta)$, k = 0,1, ..., P.
LALBT = Loc. of $a_j(\lambda,\theta)$ and $b_j(\lambda,\theta)$.

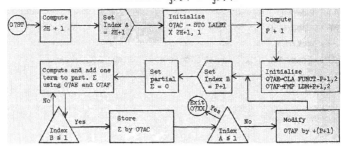

36

Compute $\Gamma(\lambda,\theta,t)$ for fixed t and varying λ and θ

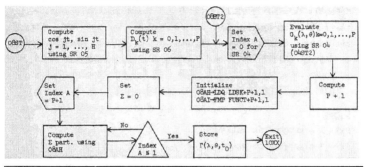

Compute $\Gamma(\lambda,\theta,t)$ for fixed λ and θ and varying t

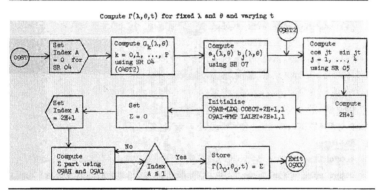

Compute $\Gamma(\lambda,\theta,t)$ for fixed $T = t-\theta$ and varying λ

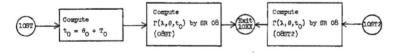

(f) SR 07 computes $a_0(\lambda, \theta)$, $a_j(\lambda, \theta)$ and $b_j(\lambda, \theta)$ (j = 1, 2, ..., H),
given the matrix of coefficients D_{sk}, the functional values
$G_k(\lambda, \theta)$ (k = 0, 1, ..., K) from SR 04, H, K, and TRA exit
instruction in 07XX. For an explanation of $a_j(\lambda, \theta)$ and $b_j(\lambda, \theta)$,
see section 3.1c of [Jones and Gallet, 1962b]. (Entrance to SR 04
is through 04ST2.)

   6.2 Program Description (for subroutines 08, 09 and 10)

(a) Input
   (1) λ = geographic latitude,
   (2) θ = geographic longitude,
   (3) t = local mean hour angle, or
        T = universal hour angle.
   (4) Matrix of coefficients D_{sk}, s = 0, 1, ..., 2H and
        k = 0, 1, ..., K.
   (5) PP1, PP2, and PP3 = P.
   (6) H = number of harmonics for the diurnal analysis.

(b) Output
   The desired functional value, $\Gamma(\lambda, \theta, t)$ or $\Gamma(\lambda, \theta, \theta+T)$ where
   t = θ + T.

(c) Storage
   Subroutine 08 and 09 each take 72 storage locations, not including
   other subroutines used. SR 10 takes 11 storage locations in
   addition to the locations used in SR 08 and other subroutines.

(d) Calling sequence
```
              C L A  EXTIN
              S T O  08XX  (or 09XX or 10XX)
              T R A  08ST  (or 08ST2 or 09ST or 09ST2 or
                                 10ST or 10ST2)
      (EXTIN)  T R A (NEXT)
      (NEXT)  (CONTINUE)
```

38

(e) Subroutines used

SR 08 uses subroutines 04, 05 and 06 (see remarks) and hence all
subroutines used by these. SR 09 uses subroutines 04, 05, and 07
(see remarks). SR 10 uses SR 08 and all subroutines used by it.

(f) Remarks

The AC and MQ overflow indicators should be off at entry to
subroutines 08, 09, and 10. They, along with their included subroutines, use index registers A, B, and C, making no provision to
restore them.

7. RESIDUAL SUBROUTINE 11

SR 11 is used to compute residuals between the original ionospheric data (A and B) and the values computed from $\Gamma(\lambda, \theta, t)$ and their root mean square. Examples of such results are given in the sample printout in appendix B. These results are used to test the "goodness of fit" of the function $\Gamma(\lambda, \theta, t)$ and to spot undue irregularities in the data. Separate results are obtained for the A and B data, since the residuals for the latter are in general systematically higher than for the former. The program description is followed by a logical diagram (flow chart 7).

(a) Input

   (1) Matrix of coefficients D_{sk},

   (2) Ionospheric data (A and B),

   (3) Station latitudes λ_i ($i = 1, 2, \ldots, N$),

   (4) Station longitudes θ_i ($i = 1, 2, \ldots, N$),

   (5) Time corrections,

   (6) Station codes,

   (7) N = number of stations (A and B data),

   (8) PP1, PP2, and PP3 = P,

   (9) H = number of harmonics for diurnal analysis, and

  (10) Exit instruction TRA, in 11XX.

(b) Output (see sample printout 2, appendix B)

   (1) For each station (A and B data) the 24 hourly residuals (original ionospheric data minus values computed from $\Gamma(\lambda, \theta, t)$,

   (2) Root mean square (RMS) of the 24 residuals from each station,

   (3) RMS of residuals from A data stations taken together,

   (4) RMS of residuals from B data stations taken together, and

   (5) RMS of residuals from all stations (A and B data) taken together.

Flow Chart 7

RESIDUAL SUBROUTINE 11

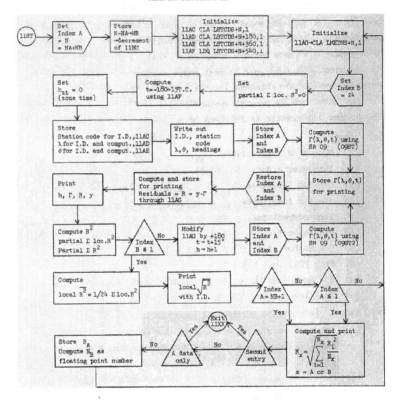

42

(c) Storage

SR 11 uses 185 storage locations (not including other subroutines employed).

(d) Calling sequence

```
              C L A  (EXTIN)
              S T O  11XX
              T R A  11ST
      (EXTIN) T R A  (NEXT)
      (NEXT)  (CONTINUE)
```

(e) Subroutines used

Share subroutine UA-SPHI, SR 09, and all included subroutines.

(f) Remarks

(1) AC and MQ overflow indicators should be off at entry to SR 11.

(2) Index registers A, B, and C are used making no provision to restore them.

(g) Stops

(1) 11BM - 4 ⎫ if an attempt is made to take the square root
(2) 11BQ - 4 ⎭ of a negative number

43

8. GENERAL DATA FITTING SUBROUTINES 15, 17-20

8.1 Explanatory Remarks

The general data fitting method described in chapters 2 and 6 of Jones and Gallet [1962a] is performed in the present program by five subroutines:

SR 20 Executive general data fitting,

SR 15 Inner products,

SR 17 Gram-Schmidt orthogonalization,

SR 19 Least squares fitting, and

SR 18 Coefficients D_{sk}.

SR 20 sets up the necessary requirements for input, output, storage, and number of entries to each of the other subroutines, and restores all index registers. Input, output, storage and calling sequences for SR 20 are included in section 8.2. The descriptions of the other subroutines are given in section 8.2e. The program logic for each of the five subroutines is given in flow chart 8.

8.2 Program Description

(a) Input (to SR 20)

(1) First of $N(K+1)$ consecutive storage locations used successively to store the functional values $G_k(\lambda_i, \theta_i)$, $A_k(\lambda_i, \theta_i)$, and $B_k(\lambda_i, \theta_i)$, $i = 1, 2, \ldots, N$, and $k = 0, 1, \ldots, K$.

(2) First of $N(2H+1)$ consecutive storage locations used for the $(2H+1)$ sets of N Fourier (time series) coefficients for geographic analysis.

(3) First of $K(K+1)/2$ consecutive storage locations used successively to store the triangular matrices of coefficients a_{kj}, b_{kj}, ℓ_{kj}, L_{kj} ($k = 1, 2, \ldots, K$ and $j = 0, 1, \ldots, k-1$).

Flow Chart 8

GENERAL DATA-FITTING SUBROUTINE 15, 17-20

SR 20   Executive General Data-Fitting

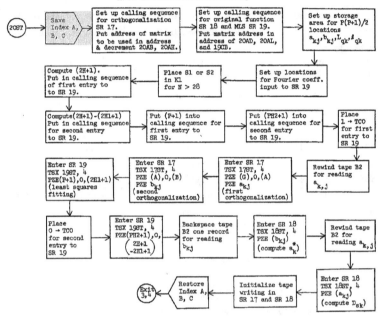

SR 15   Inner Products

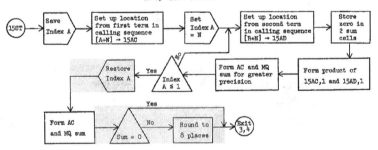

46

SR 17 Gram-Schmidt Orthogonalization

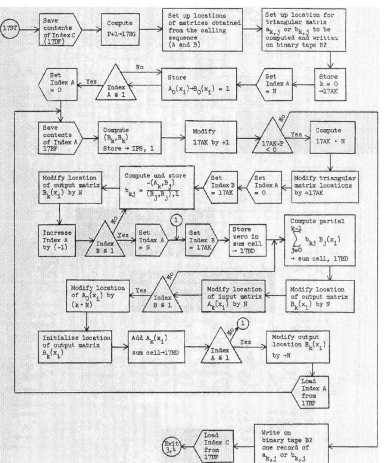

SR 19 Least Squares Fitting

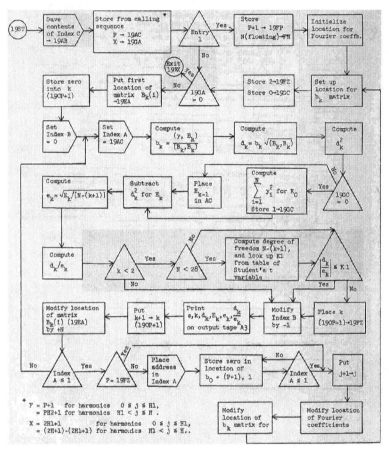

19ST → Save contents of Index C → 19AB → Store from calling sequence* F → 19AC, X → 19CA → Entry 1 — Yes → Store P+1 → 19FP, N(floating)→FN → Initialize location for Fourier coeffs.

No → Exit 19XX

Store zero into k (190P+1) ← Put first location of matrix $B_k(1)$ →19EA ← 19CA = 0 ← Store 2→19FZ, Store 0→19CC ← Set up location for b_k matrix

Set Index B = 0 → Set Index A = 19AC → Compute $b_k = \dfrac{(y, B_k)}{(B_k, B_k)}$ → Compute $d_k = b_k \sqrt{(B_k, B_k)}$ → Compute d_k^2

Compute $e_k = \sqrt{E_k / [N-(k+1)]}$ ← Subtract d_k^2 for E_k ← Place E_{k-1} in AC ← Compute $\sum_{i=1}^{N} y_i^2$ for E_0, Store 1→19CC ← Yes — 19CC = 0

Compute d_k / e_k → k < 2 — Yes → N < 28 — Yes → Compute degree of freedom N-(k+1), and look up K1 from table of Student's t variable → $\left|\dfrac{d_k}{e_k}\right| \le$ K1

Modify location of matrix $B_k(1)$ (19EA) by +N ← Put k+1 → k (190P+1) ← Print s, k, d_k, E_k, e_k, $\dfrac{d_k}{e_k}$ on output tape A3 ← Modify Index B by -1 ← Place k (190P+1)→19FZ

Index A ≤ 1 — Yes → P = 19FZ — No → Place address in Index A → Store zero in location of b_0 + (P+1), 1 ← Index A ≤ 1 — Yes → Put j+1→j

Modify location of b_k matrix for ← Modify location of Fourier coefficients

* F = P+1 for harmonics 0 ≤ j ≤ H1,
  = PH2+1 for harmonics H1 < j ≤ H .

X = 2H1+1   for harmonics 0 ≤ j ≤ H1,
  = (2H1+1)-(2H1+1) for harmonics H1 < j ≤ H..

48

SR 18 Obtaining Y_P in terms of $\Sigma\, D_k\, G_k$

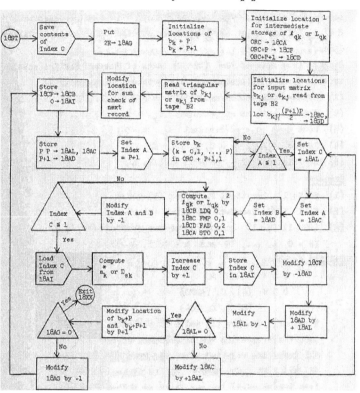

[1] $ORC+k$ = location of $\ell_{0,k}$, $k = 0,1, \ldots, P$.

$ORC-(k+1)$ = location of $\ell_{1,k}$ $k = 0,1, \ldots, P-1$.

$ORC - \sum\limits_{j=2}^{q}(P+2-\ell)-(k+1)$ = location of $\ell_{q,k}$, $k = 0,1, \ldots, P-q$, for $2 \le q \le P$.

[2] Computation starts with $\ell_{1,P-1}$ (or $L_{1,P-1}$) followed by $\ell_{1,P-2}$, $\ell_{1,P-3}$, $\ldots$, etc.

Note: The matrix ℓ_{kj} is used after second orthogonalization to transfer $\Sigma\, b_k\, B_k$ to the form $\Sigma\, a_k^*\, A_k$, see section 6.2 of Jones and Gallet [1962a]. Then the matrix L_{kj} is used to transfer $\Sigma\, a_k^*\, A_k$ to the form $\Sigma\, D_{sk}\, G_k$ (see section 2.2 [Jones and Gallet, 1962a]).

(b) <u>Output</u> (for each of the $(2H+1)$ sets of Fourier coefficients written on tape 3, channel A) (see printout in appendix B)

    (1) Orthonormal coefficients d_k, $k = 0, 1, \ldots, K$.

    (2) Sums of squares of residuals E_k, $k = 0, 1, \ldots, K$.

    (3) Root mean squared residual e_k, $k = 0, 1, \ldots, K$.

    (4) Ratios d_k/e_k, $k = 0, 1, \ldots, K$.

    (5) Coefficients D_{sk}, $k = 0, 1, \ldots, K$.

(c) <u>Storage</u>

    (1) $N(K+1)$ locations for input (1) above.

    (2) $K(K+1)/2$ location for input (3) above.

    (3) $(2H+1)(K+1)$ locations for the coefficients D_{sk}
        $(s = 0, 1, \ldots, 2H$ and $k = 0, 1, \ldots, K)$.

(d) <u>Calling sequence</u>

        T S X   2OST, 4

        P Z E (G), 0, (ADFC)

               (LFC), 0, (Kl)

Note: Kl denotes the value of Student's t variable used for truncating the orthonormal series $\Sigma\, d_k\, F_k$. All terms are rejected after the last term for which $|d_k/e_k| \geq \mathrm{Kl}$. The number Kl is either S1 or S2 (see header card) for the first or second analysis, respectively. Kl could also be the value of t for a two tail, 10 percent rejection depending upon the number of degrees of freedom

$$DF = N - (k+1),$$

Table 2

STUDENT'S t VARIABLE FOR
TWO TAIL 10% REJECTION[1]

[Degrees of Freedom (D.F.) = N - (k+1).]

D.F.	t
1	6.314
2	2.920
3	2.353
4	2.132
5	2.015
6	1.943
7	1.895
8	1.860
9	1.833
10	1.812
11	1.796
12	1.782
13	1.771
14	1.761
15	1.753
16	1.746
17	1.740
18	1.734
19	1.729
20	1.725
21	1.721
22	1.717
23	1.714
24	1.711
25	1.708
26	1.706
27	1.703
28	1.701

[1] Reference: [C.R.C., 1957].

where N is the number of Fourier coefficients being analyzed
and k+1 is the number of functions used at this point in
the analysis (see section 2.1). Table 2 contains the values
of t for a 10 percent cutoff.

A different percent can be used for the rejection criterion by
replacing the values of t in table 2 by a different set, as follows.
The program must be reassembled with the new table of values in locations DF2 to DF2 + 27. The following card format should be used.

Table 3

Card Format for Change in the Rejection Criterion

Card Column	2	8	12	18	...	72
	DF2	DEC	X.XXX,X.XXX,...,X.XXX,...			
		DEC	X.XXX,X.XXX,...,X.XXX,...			
		DEC	X.XXX,X.XXX,			
		.				
		.				

Note: No columns after 72 can be used. A blank after column
11 causes all remaining information on that card to be
ignored.

(e) Description of subroutines used

(1) SR 15 is used to form inner products of two N dimensional
vectors. Suppose U is a vector whose components are stored
in locations A, A+1, ..., A+N-1 and V is a vector with
corresponding components in B, B+1, ..., B+N-1. Then
given N and the locations A+N and B+N, SR 15 computes
the inner product (U, V) and leaves it in the accumulator.

The calling sequence is:

```
                    T S X    15ST, 4
                    P Z E    (A+N), 0, (N)
                    P Z E    (B+N)
```

52

(2) SR 17 is used to orthogonalize a set of $K+1$ linearly independent vectors of dimension N. The method employed is the Gram-Schmidt orthogonalization and reorthogonalization process described in sections 2.2 and 6.2 by Jones and Gallet, [1962a]. SR 17 forms the orthogonal functions $B_k(\lambda, \theta)$ from the geographic functions $G_k(\lambda, \theta)$ (table 1). Given the locations for input (1) and input (3), SR 17 first computes the values $A_k(\lambda_i, \theta_i)$ and stores them in place of the $G_k(\lambda_i, \theta_i)$. Following this, it computes the values $B_k(\lambda_i, \theta_i)$ (from the second orthogonalization process) and stores them in place of the $A_k(\lambda_i, \theta_i)$. The intermediate coefficients a_{kj} and b_{kj} are also computed, the b_{kj} being stored in place of the a_{kj}. (Note: in order to conserve storage space the input and output locations determined by SR 20 are the same as used here). The triangular matrices of coefficients a_{kj} and b_{kj} are written on tape 2, channel B for later use in SR 18. The same storage locations are also used by SR 18 for intermediate storage of matrices ℓ_{kj} and L_{kj}. The calling sequence is:

$$T\ S\ X \quad 17ST,\ 4$$
$$P\ Z\ E \quad (C+W),\ 0,\ (C+N)$$
$$P\ Z\ E \quad (D)$$

where C is input (1) and D is input (3) above.

(3) SR 19 is used to fit successively, by least squares, the orthogonal functions generated by SR 17 to each of the $2H+1$ sets of N Fourier coefficients (input (2)). Output from SR 19 includes output (1), (2), (3), and (4) listed above. The orthonormal coefficients d_k are set equal to zero for all k greater than the last k for which $|d_k/e_k| \geq K1$. When fitting Fourier coefficients for harmonics less than or equal H1, the calling sequence is:

$$T\ S\ X \quad 19ST,\ 4$$
$$P\ Z\ E \quad (P+1),\ 0,\ (2H1+1).$$

53

For harmonics j such that $H1 < j \leq H$, the calling sequence is:

        T S X   19ST, 4

        P Z E (PH2 + 1), 0, 2(H - H1).

(4) SR 18 is used to compute the coefficients D_{sk} (output (5) above). Input to SR 18 are the orthonormal coefficients d_k from SR 19, and the triangular matrices of coefficients a_{kj} and b_{kj} on tape 2, channel B. SR 20 determines the number of entries to SR 18 and the position of tape 2, channel B. The calling sequence for SR 18 is:

        T S X   18ST, 4

        P Z E (LGK), 0, 0

Note: location LGK which was used in SR 17 for storage of the matrices of orthogonal values is now free to be used for the storage of the triangular matrices b_{kj} and a_{kj} as they are read from tape 2, channel B.

54

9. INTERCOMPARISON COEFFICIENTS SUBROUTINES 16 AND 21

9.1 Summary of Method and Formulas

For intercomparing numerical maps of ionospheric characteristics from month to month, it is desirable to use series of orthonormal functions, since the terms in such series are independent. The orthonormal functions $F_k(\lambda, \theta)$ used for a given month (or day) depend, for their construction, on the particular set of stations available [Jones and Gallet, 1962a, chapter 2]. Therefore, since the set of available stations changes from day to day and month to month, one cannot use the orthonormal coefficients d_k for intercomparison purposes. For this reason we have made use of a special set of orthonormal functions $H_k(\lambda, \theta)$, constructed relative to a fixed set of 139 points which cover the globe approximately uniformly. The coordinates of these points are given in table 4. Thus, it is possible to represent the function $\Gamma(\lambda, \theta, t)$ in terms of a fixed set of orthonormal functions, and the resulting coefficients h_{sk} can be intercompared from one month (or day) to the next.[1]

The construction of the functions $H_k(\lambda, \theta)$ is completely analogous to the construction of the orthonormal functions $F_k(\lambda, \theta)$ [Jones and Gallet, 1962a, chapters 2 and 6]. One begins with the same set of geographic functions $G_k(\lambda, \theta)$, with the station coordinates (λ_i, θ_i) replaced by the set of 139 covering points in table 4. The first orthogonalization provides:

$$R_0 = G_0$$
$$R_k = \sum_{j=0}^{k-1} r_{kj} R_j + G_k \qquad k = 1, 2, \ldots, K \qquad (9.1)$$

---

[1]Note: The index s in h_{sk} is used here in the same sense as (56) in [Jones and Gallet, 1962a].

Table 4

COORDINATES OF THE 139 COVERING POINTS

λ = Latitude	θ = Longitude
± 90°	0°
± 80°	25° + 90°i, i = 0,1,2,3
± 70°	10° (1+6i), i = 0,1,2,3,4,5
± 10° (1+2k), k = 0,1,2	20° (1+2i), i = 0,1,...,8
± 20°k, k = 0,1,2,3	40°i, i = 0,1,...,8

where

$$r_{kj} = -\frac{(G_k, R_j)}{(R_j, R_j)} \cdot 1 \qquad\qquad (9.2)$$

A second orthogonalization yields:

$$T_0 = R_0$$

$$T_k = \sum_{j=0}^{k-1} t_{kj} T_j + R_k \qquad k = 1, 2, \ldots, K, \qquad (9.3)$$

$$t_{kj} = -\frac{(R_k, T_j)}{(T_j, T_j)} . \qquad\qquad (9.4)$$

The orthogonal functions $T_k(\lambda, \theta)$ are then normalized by

$$H_k = \frac{T_k}{\sqrt{(T_k, T_k)}} \qquad k = 0, 1, \ldots, K. \qquad (9.5)$$

Consider now the problem of transforming the expression

$$Y_k = \sum_{k=0}^{K} D_k G_k \qquad\qquad (9.6)$$

to the form

$$Y_k = \sum_{k=0}^{K} h_k H_k . \qquad\qquad (9.7)$$

The first step is:

$$Y_k = \sum_{k=0}^{K} r_k R_k , \qquad\qquad (9.8)$$

---

[1]The inner product notation (G_k, R_j) is used here to denote a summation $\Sigma\, G_k(\lambda, \theta)\, R_j(\lambda, \theta)$ over the set of 139 covering points (table 4).

which from (9.1) and (9.6) provides:

$$r_k = D_k - \sum_{j=k+1}^{K} r_{jk} D_j, \qquad k = 0, 1, \ldots, K-1, \quad \text{and}$$

$$r_K = D_K .$$
$$\text{(9.9)}$$

This gives

$$Y_k = \sum_{k=0}^{K} t_k T_k , \qquad\qquad\qquad\qquad \text{(9.10)}$$

where from (9.3) and (9.8)

$$t_k = r_k - \sum_{j=k+1}^{K} t_{jk} r_j , \qquad k = 0, 1, \ldots, K-1, \quad \text{and}$$

$$t_K = r_K = D_K .$$
$$\text{(9.11)}$$

Finally, the desired coefficients in (9.7) are obtained from

$$h_k = t_k \sqrt{(T_k, T_k)} , \qquad\qquad k = 0, 1, \ldots, K . \qquad \text{(9.12)}$$

Two subroutines are described in this section. SR 21 is an executive subroutine which employs SR 16 to compute the coefficients r_k and t_k, the numbers r_{kj}, t_{kj} and $\sqrt{(T_k, T_k)}$ being obtained from the numerical map executive program (see input 5, section 2.3). The program description is given in section 9.2.

9.2 Program Description

(a) Input

(1) Triangular matrices r_{kj} and t_{kj}, $k = 1, 2, \ldots, K$ and $j = 0, 1, \ldots, k-1$ defining orthogonal functions for the 139 covering points. (Note: these numbers are either computed by the numerical map executive program (section 2) or else read in from tape 3, channel B. See input 5 and output 5, section 2.3).

58

Flow Chart 9

INTERCOMPARISON COEFFICIENTS SUBROUTINES 16 AND 21

SR 21 Intercomparison Coefficients Executive Subroutine

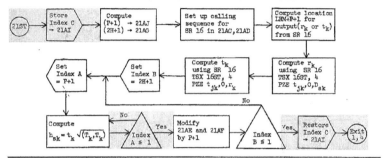

SR 16 (Computes r_k or t_k)

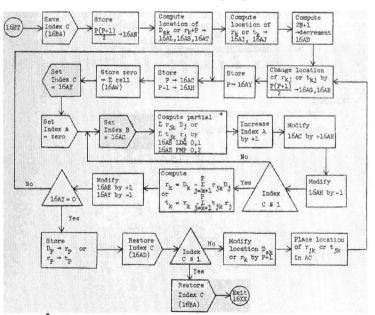

* Compute r_0 first, starting with the term $r_{PO} D_P$.

59

(2) Square roots of inner products $\sqrt{(T_k, T_k)}$, $k = 0, 1, \ldots, K$. (Note: these numbers are also either computed by the numerical map executive program (section 2) or else read in from tape 3, channel B.

(3) Matrix of coefficients D_{sk}, $k = 0, 1, \ldots, K$, $s = 0, 1, \ldots, 2H$.

(b) Output

Matrix of intercomparison coefficients h_{sk}, $s = 0, 1, \ldots, 2H$ and $k = 0, 1, \ldots, K$.

(c) Calling sequence

T S X   21ST, 4

(d) Subroutines used

SR 16 computes the coefficients r_k and t_k. The r_k are computed when entering SR 16 by

T S X   16ST, 4
(LT1), 0, 0

The t_k are computed when entering SR 16 by

T S X   16ST, 4
(LT2), 0, 0

(e) Remarks

(1) The index registers are restored after they are used.

(2) Underflow is checked internally in the code.

ACKNOWLEDGEMENTS

The authors gratefully acknowledge the assistance they
received from a number of persons. A significant contribution
to the work described here was given by Mrs. G. Anne Hessing
who was responsible for the program development using the IBM
650 computer and for the initial programming on the IBM 704.
Valuable assistance in correcting the flow charts and the program description was rendered by Mr. Ronald P. Graham. Computer
services were obtained from the operators in the Computer
Laboratories of the National Bureau of Standards (in Washington,
D. C., and in Boulder, Colorado). Able assistance in typing
the manuscript and assembling the material for this paper was
given by Mrs. Anna von Kreisler.

10. REFERENCES

C.C.I.R. (1959), Report No. 162 and Study Programme No. 149, Documents of the Plenary Assembly, Los Angeles (Published by the International Telecommunication Union, Geneva).

C.R.C. Standard Mathematical Tables (1957), 12th Ed., p. 244 (Chemical Rubber Publish. Co., Cleveland, Ohio).

Central Radio Propagation Laboratory Ionospheric Predictions (Jan. 1963), No. 1 (U. S. Dept. of Commerce, National Bureau of Standards).

Hinds, M., and W. B. Jones (Dec. 1962), Computer program for ionospheric mapping by numerical methods, Unpublished Report (National Bureau of Standards, Boulder, Colorado).

Jones, W. B., and R. M. Gallet (Dec. 1960), Ionospheric mapping by numerical methods, J. of the International Telecommunication Union, No. 12, 260-264.

Jones, W. B., and R. M. Gallet (May 1962a), The representation of diurnal and geographic variations of ionospheric data by numerical methods, Telecommunication J. 29, No. 5, 129-149; and (July-Aug. 1962a) J. Res. NBS 66D (Radio Prop.), No. 4, 419-438.

Jones, W. B., and R. M. Gallet (Nov.-Dec. 1962b), Methods for applying numerical maps of ionospheric characteristics, J. Res. NBS 66D (Radio Prop.), No. 6, 649-662.

Ostrow, S. M. (Dec. 21, 1962), Handbook for the Central Radio Propagation Laboratory ionospheric predictions based on numerical methods of mapping, National Bureau of Standards Handbook 90.

APPENDIX A

SAMPLE PRINTOUT OF INPUT TO NUMERICAL MAP PROGRAM

Header
Card
Fixed
Data

```
          3 154 52 4 8 100 12 36 52  4 -2 60 2  2 142  59 1868   4         58064500
     10687219900+868210000+066201000+790                                              5806
     10483141700+823045000+645191000+700                                              5806
     10982243900+799285000-274239000+810                                              5806
     10J82297400+826285000+083170000+850                                              5806
     10280058000+806045000+087                                                        5806
     10980274100+800285000-073235000+860                                              5806
     10178015700+782015000+005012900+744                                              5806
     10J76291300+766285000+042355000+870                                              5806
     10974265100+747270000-033289400+830                                              5806
     10373080400+735105000-164162000+640                                              5806
     10771203200+713210000-045241200+685                                              5806
     10J70291400+705285000+043000700+819                                              5806
     10169019000+697015000+027116200+669                                              5806
     10J69306500+693315000-057032700+798                                              5806
     10168033000+690030000+020126000+630                                              5806
     10167020300+678015000+035115700+652                                              5806
     10166026600+674030000-023120000+640                                              5806
     10266066700+665060000+045148000+570                                              5806
     10165022100+656015000+047114700+629                                              5806
     10164018800+646015000+025111000+630                                              5806
     10664186600+644180000+044237000+680                                              5806
     10764212200+649210000+015256600+646                                              5806
     10964264000+643270000-040315500+737                                              5806
     10A64338200+641345000-045071100+701                                              5806
     10J63291400+638285000+043360000+750                                              5806
     10462129600+620135000-036194000+510                                              5806
     10761210100+612210000+001258200+609                                              5806
     10J61314600+612315000-003036900+712                                              5806
     10160030700+600030000+005                                                        5806
     10059011100+600015000-026100000+595                                              5806
     10159024600+605030000-036113000+560                                              5806
     10158017600+598015000+017106000+585                                              5806
     10958265800+588270000-028322800+687                                              5806
     10056355800+574360000-028083400+607                                              5806
     10156044300+561045000-005126000+502                                              5806
     10256061100+567060000+007140800+484                                              5806
     10356084900+565090000-034159700+459                                              5806
     10055013400+546015000-011099000+550                                              5806
     10155037300+555030000+049120600+508                                              5806
     10855246700+546255000-055301000+620                                              5806
     10053005200+521000000+035089500+537                                              5806
     10052014600+500000000+097097500+500                                              5806
     10352104000+525105000-007174000+410                                              5806
     10452113500+520120000-043182100+405                                              5806
     10051359400+515360000-004083300+543                                              5806
     10651183400+519180000+023240100+472                                              5806
     10050010100+516015000-033093900+521                                              5806
     10049004600+501000000+031088000+520                                              5806
     10149039700+472045000-035120000+420                                              5806
     10949262600+499270000-049322900+588                                              5806
     10048007600+481000000+051090000+494                                              5806
     10848236600+484240000-023294300+533                                              5806
     10147019200+474000000+128101000+470                                              5806
     10547143000+470150000-047208500+370                                              5806
```

65

```
10J47307300+476300000+049021400+584                           5806
10046000300+466000000+002079500+499                           5806
10045007300+468015000-051088700+480                           5806
10145020500+448015000+037100000+440                           5806
10545141700+454135000+045206100+352                           5806
10945284100+454285000-006351500+569                           5806
10044009000+446015000-040                                     5806
10144034100+448030000+027113000+390                           5806
10343076900+432075000+013150500+337                           5806
10041012500+418015000-017092100+425                           5806
10040000300+408000000+002                                     5806
10J40285900+404285000+006354000+510                           5806
10539140100+397135000+034205500+294                           5806
10938282900+387285000-014350300+500                           5806
10237058300+379060000-011133000+300                           5806
10837237800+374240000-015298600+436                           5806
10535139500+357135000+030205500+254                           5806
10J34353200+309360000-045070000+380                           5806
10832253500+323255000-010317000+412                           5806
10431130600+312135000-029197900+203                           5806
10929279400+284285000-037347000+390                           5806
10328077200+286075000-015149000+188                           5806
10426127800+263135000-048195700+152                           5806
10926281800+266285000-021350000+380                           5806
10424121500+250120000+010189000+140                           5806
10.23072600+230075000-016144000+140                           5806
10022005500+228000000+037080000+260                           5806
10422113600+222120000-043278000+110                           5806
10720203500+208210000-043268200+208                           5806
10219072800+190075000-015143900+098                           5806
10J18292800+185300000-048002100+299                           5806
10416120600+164120000+004189300+050                           5806
10A14342600+148360000-116054700+215                           5806
10313080300+131075000+03515C100+030                           5806
10311078700+108075000+025148500+010                           5806
10310077500+102075000+017014800+006                           5806
10909280100+094285000-033348600+206                           5806
10308077000+085075000+013146000-014                           5806
10007003900+074000000+026074800+106                           5806
10J06304800+058360000-368015000+170                           5806
10J05285800+045285000+005345000+160                           5806
10104018600+046015000+024089000+050                           5806
10102030200+015000000+201991800-019                           5806
10301103800+013105000-008172800-101                           5806
1010K028800-023030000-008097300-038                           5806
1050K140800-025000000+939211000-120                           5806
1010M015200-044000000+101083600-029                           5806
1090M278700-046285000-042350000+066                           5806
10A0N324900-053330000-034034000+040                           5806
1090P280200-068285000-032349000+040                           5806
1090R281400-091285000-024350000+020                           5806
1011J027500-116000000+183094100+127                           5806
1091K284700-120285000-002954000-005                           5806
10J10292000-165300000-053000900-050                           5806
1071P210700-177210000+005283000-150                           5806
```

```
1021Q047500-188045000+017112500-237                                              5806
1011R017700-192015000+018083000-180                                              5806
1051R146700-193150000-022219000-284                                              5806
1072J200200-212195000+035273800-209                                              5806
10J2L313500-235315000-010022500-128                                              5806
10120028100-262030000-013091400-269                                              5806
10J20294600-269300000-036004000-160                                              5806
1052P152900-275150000+019227100-357                                              5806
10430115900-303120000-027185800-417                                              5806
1013M018300-341030000-078079700-327                                              5806
10J3M301500-345300000+010009500-231                                              5806
1053N149000-353150000-007224400-439                                              5806
10J3O287200-366285000+013356000-250                                              5806
1054K147200-429150000-019224500-516                                              5806
1064L172800-436180000-048252800-480                                              5806
10J4L294700-432300000-035003000-320                                              5806
1024R070300-494075000-031                                                        5806
10J5J302200-517300000+015009300-403                                              5806
1065K169200-525165000+028253300-572                                              5806
1055M159000-545150000+060243300-610                                              5806
10J6L299300-630315000-105006100-515                                              5806
10J6M296500-648300000-023004000-533                                              5806
1036Q093000-665090000+020146000-780                                              5806
1046O110500-662105000+037179000-780                                              5806
1056O140000-667135000+033                                                        5806
10170023300-704030000-045063000-068                                              5806
1067K170300-723165000+035277000-750                                              5806
10A7O333400-755330000+023017500-675                                              5806
10A7P318900-777315000+026147000-670                                              5806
1067Q166800-778165000+012295000-790                                              5806
1077Q197800-782195000+019312000-740                                              5806
1088024000-800240000+000336000-710                                              5806
10090000000-900000000+000000000-780                                              5806
1047112890O+716128900+000                                                        5806
10919260500+193260500+000000000+000                                              5806
10513144900+136144900+000213000+040                                              5806
10111043200+116043200+0001140Q0+069                                              5806
10411125000+110125000+000193800-001                                              5806
10705197900+059197900+000265900+051                                              5806
10701202700+019202700+000271400+022                                              5806
1050R147100-094147100+000
1061Q178200-180178200+000251200-220                                              5806
1062R182100-292182100+000257800-322                                              5806
1016R039600-690039600+000780000-700
1047P106900-784106900+000129000-880
```

Iono-
spheric
Data →

```
1168758O645000590 0600 0600 0590 0590 0550 0570 0560 0550 0550 0560 0550
1268758O645000540 0530 0500 0500 0550U0500 0510 0500 0530 0560 0560 0590
1148358O645000520 0520 0510 0520 0550 0560 0600 0580 0600 0530 0500 0550
1248358O645000560 0580 0580 0600 0590 0570 0580 0580 0570 0570 0580 0580
1198258O645000570 0560 0580 0585 0585 0560 0550 0550 0560 0540 0550 0550
1298258O645000560 0550 0570 0555 0555 0540 0570 0565 0540 0570 0560 0550
11J8258O645000530 0560 0580 0520 0540 0510 0520 0570 0580 0560 0600 0590
12J8258O645000600 0590 0600 0600 0570 0580 0560 0550 0560 0540 0540 0540
11280580645000560 0560 0550 0530 0540 0520 0550 0550 0590 0620 0650 0600
12280580645000580 0570 0600 0600 0600 0600 0600 0600 0600 0600 0600 0600
```

67

```
11980580645000550 0570 0530 0550 0550 0540 0530 0550 0550 0530 0560 0580
12980580645000580 0560 0570 0580 0580 0580 0580 0580 0550 0540 0560 0560
11178580645000510 0550 0580 0560 0570 0560 0520 0590 0620 0610 0650 0620
12178580645000620 0620 0620 0630 0630 0620 0640 0640 0630 0580 0580 0550
11J76580645000560 0575 0530 0520 0510 0530 0470E0495 0540 0520 0560 0550
12J76580645000530 0560 0575 0580 0570 0580 0595 0570 0590 0570 0550 0560
11974580645000560 0540 0560 0520 0520 0540 0540 0520 0540 0550 0580 0540
12974580645000560 0580 0540 0560 0590 0600 0600 0580 0600 0600 0600 0580
11373580645000600 0600 0570 0540 0570 0530 0560 0590 0590 0610 0620 0630
12373580645000640 0660 0660 0640 0640 0620 0620 0620 0630 0630 0620 0620
11771580645000540 0550 0520 0520 0540 0540 0550 0540 0470E0510 0590 0580
12771580645000520E0570 0600 0580 0600 0600 0590 0600 0600 0560 0560 0550
11J70580645000550 0520 0490 0520 0500 0510 0520 0530 0570 0580 0580 0600
12J70580645000570 0580 0600 0590 0570 0580 0600 0590 0580 0580 0570 0550
11169580645000560 0570 0580 0580 0640 0630 0620 0640 0640 0650 0660 0650
12169580645000640 0640 0630 0620 0630 0630 0630 0620 0590 0600 0590 0580
11J69580645000525 0520U0520U0500 0500U0485 0475U0480 0530U0560 0555U0620U
12J69580645000610U0600 0590U0560U0580 0575U0580 0610U0590U0590 0555 0520
11168580645000620 0620 0630 0620 0620 0640 0630 0650 0640 0660 0680 0680
12168580645000660 0660 0650 0630 0630 0640 0640 0640 0620 0620 0640 0600
11167580645000560 0600 0590 0610 0610 0600 0600 0610 0630 0640 0650 0630
12167580645000630 0630 0630 0620 0630 0640 0630 0620 0610 0600 0600 0600
11166580645000640 0620U0630U0600U0660 0640 0630 0650 0650 0690 0680 0660
12166580645000660 0650 0650 0650 0650 0650 0650 0660 0660 0650 0600 0620
11266580645000600 0620 0640 0640 0660 0670 0640 0670 0690 0680 0700
12266580645000700 0710 0700 0680 0660 0670 0670 0680 0660 0650 0660 0620
11165580645000620 0620 0610 0640 0650 0630 0640 0670 0680 0690 0670 0670
12165580645000670 0660 0660 0650 0650 0660 0660 0650 0660 0640 0620 0630
11164580645000630 0630 0630 0630 0640 0640 0640 0640 0680 0670 0680 0670
12164580645000670 0660 0660 0650 0670 0660 0670 0650 0670 0650 0620 0630
11664580645000650U0630U0620 0640 0660 0660 0640 0680 0610 0620 0610 0600
12664580645000600 0620 0620 0630 0630 0640 0630 0640 0640 0620 0650
11764580645000530U0550U0540 0590 0560U0600U0580 0580 0600 0620 0595 0595
12764580645000610 0600 0610 0605 0600 0605 0600 0570 0580 0615 0550 0530U
11964580645000560 0540 0530 0540 0510 0530 0530 0550 0570 0580 0600 0600
12964580645000630 0680 0680 0660 0660C 0640 0650 0620 0600 0610 0620 0610
11A64580645000500D0540U0520U0530U0530 0560 0560 0560 0580 0600 0620 0630
12A64580645000640 0650 0670 0650 0640 0630 0630 0590 0580D0580U0550U0530U
11J63580645000530 0500 0520 0520 0480 0460 0500 0450 0480 0500 0500 0620
12J63580645000660 0640 0640 0650 0640 0620 0630 0600 0560 0600 0560 0540
11462580645000820 0780 0700 0720 0710 0720 0700 0720 0720 0720R0720R0720R
12462580645000730R0730R0740R0740R0760U0740 0740 0760 0760 0780 0800 0800
11761580645000530 0520 0540 0550 0580 0610 0640 0630 0640 0620 0600 0600
12761580645000600 0620 0620 0610 0600 0600 0620 0600 0610 0600 0590 0540
11J61580645000480U0500U0470U0465U0490 0470 0520 0570 0585 0620 0630 0640
12J61580645000655 0680 0670 0650 0640 0630 0630 0610 0590 0560U0550U0520
11160580645000700 0700 0700 0710 0760 0760 0730 0760 0760 0790 0780 0780
12160580645000720 0700 0760 0700 0750 0760 0710 0700 0710 0700 0730 0710
11059580645000700 0690 0660 0660 0670 0660 0660 0680 0670 0680 0690 0700
12059580645000690 0690 0690 0680 0690 0720 0700 0700 0690 0680 0650 0640
11159580645000700 0690 0690 0700 0680 0710 0680 0670 0710 0710 0720 0730
12159580645000710 0700 0690 0690 0680 0680 0700 0700 0700 0700 0700 0700
11158580645000680 0640 0650 0660 0670 0670 0680 0690 0690 0710 0720 0700
12158580645000700 0700 0680 0690 0700 0700 0700 0700 0700 0700 0700 0700
11958580645000560 0520 0500 0500 0500 0490 0500 0530 0570 0600 0620 0630
```

```
12958580645000650 0690 0720 0700 0690 0670 0640 0610 0600 0580 0570 0560
11056580645000700 0680 0680 0680 0700 0670 0710 0680 0690 0690 0700 0730
12056580645000720 0700 0700 0730 0740 0760 0760 0730 0730 0710 0720 0720
11156580645000700 0690 0680 0680 0690 0690 0740 0760 0740 0750 0790 0790
12156580645000790 0760 0740 0730 0700 0700 0700 0700 0690 0700 0700 0710
11256580645000720 0680 0660 0640 0670 0700 0730 0730 0730 0750 0780 0800
12256580645000800 0800 0780 0750 0740 0730 0720 0710 0700 0720 0730 0730
11356580645000740 0710 0690 0670 0680 0700 0720 0720 0730 0750 0770 0760
12356580645000790 0810 0800 0780 0760 0740 0740 0750 0720 0740 0770 0770
11055580645000770 0700 0690 0680 0700 0720 0710 0730 0740 0730 0730 0770
12055580645000760 0740 0750 0730 0740 0750 0780 0780 0780 0770 0770 0780
11155580645000710 0680 0650 0660 0690 0740 0750 0740 0750 0770 0780 0780
12155580645000770 0760 0740 0720 0710 0720 0740 0720 0730 0740 0740 0740
11855580645000520 0500 0500 0490 0500 0530 0580 0610 0610 0610 0630 0630
12855580645000640 0650 0670 0640 0650 0650 0670 0680 0680 0630 0560 0570
11053580645000700 0690 0650 0640 0690 0700 0720 0740 0790 0760 0770 0730
12053580645000760 0760 0740 0730 0730 0760 0750 0780 0770 0760 0760 0740
11052580645000690 0650 0660 0650 0700 0760 0780 0850 0830 0800 0830 0800
12052580645000810 0800 0800 0780 0800 0780 0800 0790 0780 0750 0720 0700
11352580645000780 0740 0720 0690 0680 0720 0750 0800 0770 0740 0780 0800
12352580645000790 0790 0810 0790 0780 0770 0750 0780 0790 0820 0820 0810
11452580645000820 0780 0730 0690 0700 0760 0760 0770 0780 0760 0740 0750
12452580645000760 0770 0770 0780 0780 0780 0770 0790 0820 0820 0820
11051580645000740 0720 0690 0660 0680 0710 0740 0780 0740 0780 0760 0780
12051580645000800 0780 0770 0780 0780 0810 0800 0820 0800 0800 0800 0820
11651580645000660 0620 0600 0590 0590 0660 0710 0740 0710 0680 0690 0690
12651580645000680 0660 0650 0650 0630 0650 0670 0680 0690 0680 0700 0680
11050580645000765 0706 0690 0660 0655 0707 0730 0756 0786 0785 0758 0775
12050580645000778 0768 0769 0764 0760 0768 0779 0824 0796 0798 0793 0792
11049580645000700 0680 0660 0630 0650 0690 0720 0750 0750 0760 0770 0770
12049580645000770 0760 0750 0760 0750 0760 0800 0790U0790 0770 0770 0750
11149580645000740 0740 0720 0660 0660 0710 0750 0760 0790 0820 0900 0910
12149580645000920 0900 0860 0830 0810 0800 0790 0820 0800 0800 0770 0770
11949580645000570 0510 0500 0480 0460 0500 0540 0560 0600 0600 0620 0630
12949580645000650 0660 0650 0680 0680 0680 0700 0700 0700 0700 0690 0600
11048580645000700 0690 0660 0640 0680 0750 0790 0810 0800 0800 0810 0880
12048580645000810 0830 0800 0800 0780 0800 0840 0810 0830 0820 0820 0770
11848580645000520 0480 0450 0420 0430 0510 0570 0590 0620 0610 0630 0650
12848580645000660 0660 0670 0650 0650 0640 0650 0650 0650 0630 0600 0540
11147580645000720 0690 0670 0680 0740 0790 0770 0860 0900 0900 0870 0870
12147580645000850 0840 0850 0840 0850 0820 0830 0850 0750 0690 0750 0750
11547580645000880 0880 0840 0800 0780 0820 0880 0900 0900 0870 0820 0820
12547580645000840 0840 0820 0820 0820 0820 0820 0820 0860 0860 0880 0910
11J47580645000690 0620 0590 0520 0520 0560 0600 0650 0680 0690 0700 0720
12J47580645000700 0720 0740 0760 0760 0760 0800 0800 0800 0820 0780 0710
11046580645000800U0780U0720U0680U0680 0740 0800U0800 0800 0830 0870 0870
12046580645000880 0870 0860 0860 0860 0850 0860U0870U0880 0850D0860U0860
11045580645000790 0720 0700 0670 0650 0670 0740 0790 0810 0850 0820 0850
12045580645000830 0850 0820 0840 0800 0790 0790 0820 0800 0790 0790 0780
11145580645000770 0760 0745 0713 0720 0803 0875 0868 0885 0922 0932 0946
12145580645000900 0930 0912 0874 0876 0878 0873 0871 0834 0823 0802 0792
11545580645000840 0820 0780 0750 0760 0840 0880 0880 0850 0810 0760 0800
12545580645000800 0760 0780 0800 0780 0800 0780 0810 0830 0830 0840 0850
11945580645000540 0560 0500 0490 0460 0510 0550 0570 0600 0620 0660 0680
12945580645000680 0690 0700 0710 0720 0720 0750 0800 0800 0800 0730 0660
```

```
1104458064500072O 0720 0710 0700 0670 0680 0750 0800 0880 0850 0920 0930
1204458064500094O 0910 0900 0870 0870 0860 0850 0840U0730 0690 0720 0730
1114458064500080OU0760U0720U0700U0700U0780 0860 0900 0880 0900 0920 0950
1214458064500095O 0910 0880 0880 0840 0830 0860 0840 0840U0830U0820 0820U
1134358064500079O 0770 0740 0700 0690 0780 0880 0920 0970 1020 1020 1030
1234358064500104O 1010 1000 0960 0910 0870 0860 0850 0840 0830 0830 0820
1104158064500085O 0800 0810 0770 0700 0740 0860 0860 0900 0860 0940 0980
1204158064500100O 0950 0970 0930 0890 0880 0880 0900 0870 0870 0860 0880
1104058064500083O 0820 0780 0760 0740 0790 0830 0860 0910 0920 0930 0940
1204058064500094O 0930 0940 0930 0900 0880 0880 0880 0850 0840 0850 0840
11J4058064500068O 0640 0620 0580 0540 0560 0605 0620 0640 0640 0670 0660
12J4058064500068O 0700 0710 0710 0740 0760 0780 0800 0820 0810 0785 0750
1153958064500092O 0910 0850 0800 0800 0890 0950 0980 0940 0930 0910 0930
1253958064500093O 0930 0930 0920 0900 0880 0870 0860 0860 0870 0900 0910
1193858064500070O 0650 0620 0560 0530 0550 0600 0620 0670 0690 0690 0690
1293858064500071O 0710 0720 0720 0730 0740 0760 0760 0800 0800 0770 0740
1123758064500820R0810R0780 0740 0700 0750 0860 0940 0970R1000R1040 1070R
1223758064500108O 1080R1080 1040R0990R0940R0910R0870R0860 0840R0840R0830
1183758064500063O 0600 0590 0570 0550 0550 0640 0680 0710 0700 0705 0750
1283758064500078O 0780 0790 0770 0750 0750 0720 0700 0680 0670 0650 0620
1153558064500094O 0920 0860 0830 0800 0860 0940 0970 0960 0960 0970 1010
1253558064500102O 1030 1020 1000 0970 0940 0920 0890 0840 0880 0920 0940
1103458064500095OU0950 0910U0900 0830 0770 0800 0900 0900 0900 0920 1000
1203458064500103O 1040 1030 1020 1020 0960 1000 0980 0900 0900U0920U0950U
1183258064500067O 0660 0660 0610 0600 0575 0650 0710 0770 0800 0850 0880
1283258064500085O 0885D0880D0865 0800 0795 0780 0770 0770 0730 0700 0670
1143158064500098O 0900 0860 0830 0830 0900 0930 0920 0930 0960 1040
1243158064500106O 1100 1100 1130 1120 1120 1060 1000 0940 0900 0940 0970
1192958064500082O 0800 0740 0700 0690 0660 0680 0770 0780 0870 0850 0860
1292958064500090O 0920 0900 0890 0880 0860 0860 0850 0850 0820 0800 0800
1132858064500094OR0920R0880R0840R0800R0820R0900 0950 0990R1020 1080 1160
1232858064500124OR1300R1320 1320R1310R1270R1190R1110R1030R0980R0960R0950R
1142658064500107O 1045 0950 0895 0840 0785 0830 0865 0870 0910 0970 1040
1242658064500111O 1150 1210 1260 1275 1265 1210 1150 1080 1000D1020 1080
1192658064500086O 0830 0780 0760 0725 0685 0735 0800 0840 0875 0925 0950
1292658064500095O 0960 0950 0930 0935 0900 0915 0890 0880 0860 0840 0830
1142458064500124O 1120 0970 0910 0820 0820 0880 0900 0940 1000 1120 1210
1242458064500125O 1310 1380 1410 1400 1400 1360 1280 1250 1240 1160 1160
1122358064500090OR0840 0760R0710R0700 0710 0780 0930 0950 1000 1080 1220
1222358064500130OR1400R1480R1530R1500R1440R1380R1290R1160 1100 0960 0920
1102258064500120OD1090R0950 0820D0750 0820 0900 0930 0950 1060 1150 1240
1202258064500135O 1400 1440 1470R1490 1470R1420D1360R1300D1310D1280R1240D
1142258064500130O 1280 1120 0990 0810 0730 0800 0880 0920 1000 1080 1200
1242258064500127O 1400 1400 1450 1450 1400 1400 1350 1240 1150 1150 1220
1172058064500088O 0850 0785 0745 0715 0680 0680 0770 0850 0930 0990 1070
1272058064500112O 1155 1170 1225 1260 1210 1145 1100 1065 1020 0930 0910
1121958064500097OR0940R0890R0830R0770R0690R0760R0970R1010I1040 1120 1210
1221958064500129O 1290 1340 1360R1360R1350R1330R1290R1240R1180R1110R1050R
11J1858064500104O 0990 0930 0900 0860 0800 0780 0870 0920 1000 1060 1120
12J1858064500118O 1190 1180 1140 1120 1080 1060 1060 1040 1060 1060 1060
1141658064500111O 1050 0960 0905 0800 0685 0810 0900 0990 1060 1150 1200
1241658064500125O 1300 1320 1305 1295 1250 1235 1195 1100 1100 1100 1125U
11A4590645000740R0720R0680R0630R0590R0570R0550R0700 0860 0950 1050 1140
12A1458064500122O 1290R1350R1380R1390R1370R1330R1240U1110U0980R0870R0790R
1131358064500090OU0850U0830U0750U0700R0730R0910U1090R1150 1220 1150 1150
```

```
1231358064500114O 116O 117O 118OU118OU117OU11OOU1O5OUO95OUO9OOUO97OUO9OOU
1131158064500099ORO94ORO86ORO77ORO68ORO66O O93O 112O 117O 12OO 116O 112O
1231158064500111O 111O 116O 116O 118O 118OR116OR11OOR1O2OR1O5OR1O5OR1O3OR
1131058064500092ORO9OORO87ORO83ORO72O O6OO O88O 11OO 115O 119O 12OO 11OO
1231058064500108O 109O 111O 112O 114O 116O 116O 106O O94O O91ORO94ORO96OR
119O958064500096O O9OO O88O O81O O755 O69O O67O O765 O87O O97O 1O2O 113O
129O958064500118O 12OO 123O 122O 118O 114O 11OO 103O O97O 1OOO O99O O99O
1130858064500101ORO 98ORO94ORO89ORO81ORO66ORO87OR113O 123O 127OR122OR113O
1230858064500111OR114O 115O 113OR115OR116OR115OR11OORO98ORO98OD1OOOD1O2OD
1100758064500052OJO52OJO56OJO57OJO5OOUO4OO O84O 113O 129O 136O 133O 123O
1200758064500116O 112O 112O 111O 112O 111O 1O7OUO91OUO74OUO68OJO65OJO56OJ
11JO6580645001160 12OO 12OO 119O 1O2O O98O O96O O88O O82O O72O O8OO O9OO
12JO658064500100O 109O 119O 12OO 126O 127O 126O 12OO 117O 113O 11OO 106O
11JO558064500089O O85O O8OO O79O O7OO O59O O63O O795 O895 O99O 1O8O 1165
12JO558064500121O 127O 132O 1345 1295 126O 1215 115O 11OO 113O 1135 1OOO
1110458064500117ORO8OR093ORO77ORO64ORO54O O86O 125O 138O 144O 142O 14OO
121O458064500140O 131O 125O 128O 124O 125O 12OO 115ORO111OR112OR114OR116OR
1110258064500112OR0950R0840RO71ORO86OR126OR141OR142OR142OR142O 14OO 136O
121O258064500134OR132O 134O 134O 134ORO132OR132OR124OR118OR115OR112OR1O9OR
113O158064500123OR11OOR095OR081ORO69O O56O O7OO 114O 136O 15OO 151O 148O
123O158064500140OR136OR131O 128O 128O 129O 131OR132O 135OR136OR135OR131OR
1110K58064500156OR14OOR118OR1O3ORO86ORO73ORO78O 122OR142OR139OR137O 141O
121OK580645001460 143O 139O 141O 143O 145OR149OR154OR159OR163OR164OR162OR
1150K58064500132O 133O 136OU134O 135OU132O 134O 132O 137O 13OO 136O 132O
125OK58064500135O 135O 126O 116O O9OO O84O O8OO O73O O65O O98O 13OO 135O
1110M58064500127OR1O4ORO9OORO72ORO49ORO67OR111OR128O 13OO 134O 133O 136O
121OM58064500140O 141O 146O 144O 145O 151O 158OR162OR164OR163OR159OR152O
1190M58064500094O O96O O96O O895 O74O O615 O52O O75O O875 O965 1O1O 1O4O
1290M58064500106O 108O 109O 1O7O 1O7O 1O6O 1O4O 1OO5 O99O 1OOO 1O15 O995
11AON58064500051ORO5OORO49ORO48OR047CRO46ORO45ORO84OU112OU126OR132OU131OU
12AON58064500130OU125OU12OOU121OR121OR12OOR11OORO91ORO78ORO63ORO56ORO52OR
1190P58064500086O O88O O885 O82O O7OO O595 O56O O76O O89O O98O 1O3O 1O2O
1290P58064500103O 1O5O 1O35 1O1O 1OOO 1O1O O955 O9OODO86O O895DO895DO86OD
1190R58064500082O O84O O82O O725 O63O O59O O55O O78O O925 1O5O 1O5O 1O4O
1290R58064500102O 1O25 1O3O 1OOO O99O O985 O92O O865 O85O O85O O84O O82O
1111J58064500056O O4OO O35O O31O O41O O9OO 119O 125O 122O 125O 12OO 118O
1211J58064500118O 117O 116O 117O 122O 118O 113O 1O4O 1O6O O9OO O86O O65O
1191K58064500077O O78O O71O O65O O57O O53O O55O O835 1O45 11OO 11OO 1O7O
1231K58064500105O 1O35 1OOO O97O O96O O95O O9OO O84O O82S O81O O8OO O77O
11J1O58064500080O O79O O765 O69O O57O O515 O465 O66O 1OOO 1195 126O 121OD
12J1O58064500120OU116OU114O 1O8O 1O3O 1O4O O96O O9OO O89O O875 O86O O8OO
1171P58064500110O O96O O8OO O68O O52O O5OO O61O 1O3O 139O 14OO 138O 14OO
1271P58064500133O 136O 136O 135O 142O 15OO 159O 17OOD17OOD17OOD143O 114O
1121Q58064500034O O31O O29O O27O O24O O26O O42OUO9OO 115O 123O 12OO 12OO
1221Q58064500116O 112O 111O 11OO 1O6O 1O7O O94O O78O O6OO O56O O51OUO42O
1111R58064500036O O324 O3O2 O276 O25O O247 O426 O85O 113O 12O8 122O 121O
1211R58064500117B 1177 1161 1166 1159 1165 1O65 O854 O7O7 O616 O489 O43O
1151R58064500061ORO58ORO53ORO47ORO42ORO4OORO42O O84ORI19OR136OR132OU126OR
1251R58064500125OR123OR122OR12OO 116OR1O6ORO96ORO85OR078OR076OR072OR067OR
1172J58064500070O O62ORO55ORO48ORO42ORO44O O65O 1O9OU135O 144O 133O 129O
1272J58064500126O 127O 13OO 13OO 128OU132OU134OR129OR118OR1O5ORO92ORO8OOR
11J2L58064500093O O88O O83O O66O O55O O47O O42O O72O 1O2O 122O 134O 13OO
12J2L58064500131O 134O 134O 138O 14OO 142OU134OU113O 114O 112C 111O O98O
1112O58064500032O O32O O31O O3OO O3OO O28O O3OO O66O O99O 118O 125O 12OO
1212O58064500120O 118O 118O 114O 116O 11OO 1O1O O78O O62O O5OO O38O O33O
```

```
11J20580645001050R0950R0920 0740R0600R0450R0420 0580R1080U1270R1410R1390R
12J20580645001390R1420R1430R1450R1460R1440R1350R1290R1260R1270R1270R1170R
1152P580645000530 0480 0480 0490 0460 0420 0440 0820 1060 1170 1200 1180
1252P580645001180 1160 1120 1120 1070 1050 0890 0760 0660 0630 0560 0500
1143058064500430 0420 0420 0420U0420 0400 0400R0500R0850D1070R1190R1210R
12430580645001220R1220R1210R1210R1190R1130R1000R0820R0700R0580R0510R0450R
1113M580645000280 0270 0270 0280 0290 0260 0260 0270 0680 0930 1050 1140
1213M580645001240 1220 1220 1180 1210 1180 1030 0810 0570 0360 0310 0260
11J3M580645000770R0740R0680R0610R0540 0440R0370R0680R1000 1120 1200 1200R
12J3M580645001180R1180R1230R1220R1260 1130R1050R1070R1040R0960R0900R0840R
1153N580645000500U0490U0480 0490 0480 0430 0420 0580U0940U1100 1180 1230U
1253N580645001200 1200 1200 1180 1150 1100 0900D0780D0720U0590U0520U0500U
11J30580645000600 0560 0540 0500 0470 0410 0440 0710 1080 1180 1190 1160
12J30580645001160 1220 1290 1230 1160 1120 0960 0930 0840 0780 0690 0640
1154K580645000420U0420 0400R0410 0390 0370R0360 0370R0700D0950R1100R1170R
1254K580645001200R1200D1200D1180R1120R1030R0900D0760D0620R0520U0460U0440R
1164L580645000530 0490 0470 0460 0470 0470 0430 0400 0640 0930R1080R1180R
1264L580645001220R1220R1190R1160R1140 1060 0890R0790 0700 0590 0560 0530
11J4L580645000580D0570D0550R0530R0510 0490R0460R0470 0800R1000D1100D1100D
12J4L580645001100D1100D1100D1090D1060D1000D0890R0790R0700R0660R0620R0590
1124R580645000270R0270R0260R0250R0240 0230 0260 0300 0500U0780 0900 1050
1224R580645001210 1300 1400 1300 1190U1030R0850R0640R0480R0350 0290R0270R
11J5J580645000030 0330 0320 0320 0330 0320 0280 0380 0680 0910 1010 1080
12J5J580645001070 1000 0950 0920 0740 0580 0470 0380 0320 0300 0310 0330
1165K580645000460 0460 0440 0440 0440 0410 0360 0400 0650 0900 1050 1150
1265K580645001220 1160 1210 1120R1000 0820R0690R0600 0570 0550 0490 0470
1155M580645000410R0410R0420R0420R0410R0420R0390R0450R0660R0840R1000R1120R
1255M580645001180R1190R1150R1030R0830R0650R0560R0500R0460R0440R0430R0420R
11J6L580645000340 0330 0330 0340 0340 0360 0330 0340 0330 0440 0700 0890
12J6L580645000990 1010 1020 0930 0820 0710 0560 0400 0340 0330U0330U0350
11J6M580645000230 0230 0230 0230 0220 0220 0220 0230 0310 0550U0750 0880
12J6M580645000960 0980 0840 0740 0660 0520 0340 0240 0210 0230 0220 0230
11360580645000410 0340 0350 0340 0360 0390 0380 0380 0420 0520R0630R0710
12360580645000760R0780R0750 0700R0650R0610 0600 0570R0520R0470 0410 0390
11460580645000400R0390R0380R0370R0360R0370R0380R0410R0470R0600R0750R0810R
12460580645000830R0820R0790R0720R0690R0700R0670R0640R0590R0530R0470R0420R
11560580645000450U0410U0410U0360U0360U0360U0340U0340U0440R0580 0660U0680U
12560580645000710U0730U0750U0750R0740R0730U0670R0600R0530R0490R0470R0460R
11170580645000200U0260U0270U0300 0300U0320U0340 0420U0420U0450U0590
12170580645000710 0760 0870 0820 0800 0710 0500 0340 0240 0210 0210 0200
1167K580645000440 0440 0420 0390 0370 0380U0420 0420 0490 0480U0510U0680
1267K580645000720 0600 0660 0740 0700 0760 0840 0720 0700 0640 0530 0600
11A70580645000200 0240R0270 0320R0370 0400 0400 0420 0370 0370U0400 0480
12A70580645000580U0580 0530U0440 0380 0300 0220 0190 0170 0190 0200 0200
11A7P580645000370R0410R0440R0460R0470R0470R0440U0410R0370U0350U0380U0400U
12A7P580645000460 0480 0450U0375U0340U0260U0230 0230R0230R0250R0280R0320R
1167Q580645000500 0460 0440 0470 0430U0450 0410 0440 0440 0480 0560 0550
1267Q580645000520 0630 0680 0710 0810 0800 0760 0760 0640 0600 0540 0540
1177Q580645000520U0500U0490U0430U0430U0430U0400U0385U0380U0335U0390U0400U
1277Q580645000410U0470U0450U0500U0550U0600U0640U0650U0580U0500U0550U0500U
11880580645000620U0630R0610R0600R0580R0510R0440R0350 0320R0300U0320R0340R
12880580645000370R0360R0370R0380R0420R0430R0430R0500R0560U0620R0620U0620R
11090580645000480U0520U0510U0520U0600U0540U0510U0520U0530U0530U0520U0520U
12090580645000500U0490U0520U0490U0500U0550U0470U0440U0450U04700480U0500U
11471580645000640P0640P0640P0630P0610P0590P0580P0580P0580P0580P0590P0600P
```

```
12471580645000600P0610P0620P0640P0650P0660P0670P0670P0660P0650P0650P0640P
11919580645000960P0910P0850P0810P0760P0700P0760P0860P0940P1020P1080P1140P
12919580645001190P1230P1240P1220P1190P1160P1130P1090P1040P1000P0980P0960P
11513580645000970P0930P0890P0830P0760P0700P0830P0940P0990P1030P1090P1150P
12513580645001190P1230P1260P1290P1310P1330P1300P1230P1150P1080P1040P1010P
11111580645000800P0760P0720P0680P0630P0570P0790P0980P1090P1130P1150P1140P
12111580645001150P1170P1200P1220P1240P1240P1210P1150P1050P0960P0890P0850P
11411580645000990P0950P0890P0800P0700P0610P0840P1030P1120P1160P1160P1160P
12411580645001180P1190P1190P1190P1190P1190P1180P1120P1030P0990P0990P1000P
11705580645001160P1110P1040P0950P0810P0630P0710P0890P0990P1060P1100P1130P
12705580645001150P1170P1200P1210P1200P1180P1150P1080P1050P1080P1130P1160P
11701580645001090P1050P0990P0900P0780P0590P0700P0870P0990P1050P1090P1120P
12701580645001130P1150P1160P1180P1170P1140P1090P1050P1020P1030P1070P1100P
11501580645000850P0750P0600P0460P0370P0320P0530P0960P1190P1270P1230P1190P
12501580645001160P1160P1130P1080P1060P1040P1040P1050P1010P0920P0880P0860P
11610580645000770P0690P0580P0510P0440P0400P0530P1040P1280P1360P1370P1360P
12610580645001340P1310P1290P1270P1280P1260P1190P1080P0960P0900P0870P0830P
11625580645000520P0520P0520P0520P0510P0480P0450P0730P1000P1140P1210P1180P
12625580645001150P1130P1110P1090P1060P0990P0850P0730P0640P0580P0540P0530P
11165580645000400P0400P0390P0380P0380P0400P0430P0450P0460P0470P0560P0710P
12165580645000810P0880P0890P0820P0730P0610P0470P0360P0340P0360P0380P0390P
11475580645000340P0330P0310P0310P0320P0350P0380P0420P0460P0510P0560P0620P
1247P580645000660P0670P0660P0630P0580P0550P0510P0490P0450P0420P0380P0360P
 93000 80337 65157 40210 40330 35030 35105 35185 35310 30165 25050 25325 20030
 20232 15175 10240 05160 05345 00060 00180 00225 00255-05080-05125-05200-10165
-10350-15105-15240-20075-20265-25345-30095-30225-35050-35075-35200-35265-40345
-45095-45120-45240-45325-50015-50045-50200-50270-55105-60000-60330-65210-65240
-65265-75067-75277-80037-80300-85140-85225
```

Screen
Point
Coordi-
nates

73

APPENDIX B

SAMPLE PRINTOUT OF OUTPUT FROM NUMERICAL MAP PROGRAM

DATE=5806 DATA=4500 N=154 P=52 H=8 FN=100
PP1= 12 PP2= 36 PP3= 52 PH2= 4 H1=4 TA=2 S1= 6.0 S2= 2.0
NA=142 NC= 59 S=186.8 MN=4

First Analysis

s	k	$\bar{d}_k$ ORTHO COEFF.	E_k SUMS EK	e_k RMS (EK)	$\bar{d}_k/e_k$ ORTHO COEFF/RMS
0	0	9.7796171E 01	6.6465771E 02	2.0842668E 00	4.6921139E 01
0	1	-7.9743505E-01	6.6402180E 02	2.0901111E 00	-3.8152758E-01
0	2	-2.3183237E 01	1.2655933E 02	9.1550053E-01	-2.5323018E 01
0	3	-7.2388791E-01	1.2603531E 02	9.1664355E-01	-7.8971581E-01
0	4	1.4358329E 00	1.2397369E 02	9.1216128E-01	1.5740999E 00
0	5	1.2821402E 00	1.22329B1E 02	9.0914947E-01	1.4102634E 00
0	6	-1.1086683E-01	1.2231752E 02	9.1219074E-01	-1.2153909E-01
0	7	-1.1094761E 00	1.2108658E 02	9.1069211E-01	-1.2182779E 00
0	8	5.1708430E-01	1.2081920E 02	9.1281754E-01	5.6647279E-01
0	9	-2.9567594E-01	1.2071178E 02	9.1565011E-01	-3.2291367E-01
0	10	1.9614815E-01	1.2069330E 02	9.1869968E-01	2.1350628E-01
0	11	9.0033002E-01	1.1988271E 02	9.1882774E-01	9.7986813E-01
0	12	1.1211874E 00	1.1862565E 02	9.1723313E-01	1.2223582E 00
0	13	-3.1900162E-02	1.1862463E 02	9.2049918E-01	-3.4655285E-02
0	14	4.2415078E 00	1.0063424F 02	8.5087444E-01	4.9848809E 00
0	15	1.8129360E 00	9.7347503E 01	8.3989086E-01	2.1585376E 00
0	16	4.5197838E 00	7.6919057E 01	7.4930165E-01	6.0319949E 00
0	17	-1.3756838E 00	7.5026000E 01	7.4273936E-01	-1.8524449E 00
0	18	-1.4438721E-01	7.5005152E 01	7.4538158E-01	-1.9370911E-01
0	19	4.2664412E-01	7.4823126E 01	7.4724931E-01	5.7095284E-01
0	20	-2.0292717E 00	7.0705182E 01	7.2912132E-01	-2.7831743E 00
0	21	-8.5872000E-01	6.9967782E 01	7.2805147E-01	-1.1794771E 00
0	22	1.8782378E 00	6.6440004E 01	7.1216262F-01	2.6373720E 00
0	23	9.8311927E-02	5.6430339E 01	7.1484445E-01	1.3752912E-01
0	24	-2.1768287E 00	6.1691756E 01	6.9154223E-01	-3.1477884E 00
0	25	-6.1193166E-01	6.1317295E 01	6.9212813E-01	-8.8413059E-01
0	26	-1.3842770F 00	5.9401072E 01	6.8390421E-01	-2.0240802E 00
0	27	6.5989627E-01	5.8966928E 01	6.8409902E-01	9.6315920E-01
0	28	3.2180943E 00	4.8610796E 01	6.2360753E-01	5.1604480E 00
0	29	4.1557774E-01	4.8438091E 01	6.2500380E-01	6.6492034E-01
0	30	7.0715965E-01	4.7938016E 01	6.2429157E-01	1.1327394E 00
0	31	2.0153300E 00	4.7897401E 01	6.2657931E-01	3.2164004E-01
0	32	-3.1065224E 00	3.8246920E 01	5.6221903E-01	-5.5254663E 00
0	33	-8.8725764E-01	3.7459693E 01	5.5871648E-01	-1.5880284E 00
0	34	-6.5550982E-01	3.7030000F 01	5.5783193E-01	-1.1751027E 00
0	35	1.6988062E 00	3.4144057E 01	5.3791859E-01	3.1581103E 00
0	36	1.9383783E 00	3.0346747E 01	5.0962317E-01	3.8035522E 00
0	37	5.3007979E-02	3.0343936E 01	5.1179144E-01	1.0357340E-01
0	38	1.4382064E-01	3.0363252E 01	5.1383681E-01	2.7989555E-01
0	39	1.4322262E 00	2.8311980E 01	4.9834797E-01	2.8739480E 00
0	40	-1.2305239E 00	2.6797791E 01	4.8697904E-01	-2.5268517E 00
0	41	-6.1052668E-01	2.6425048E 01	4.8573442E-01	-1.2569146E 00
0	42	-5.9274504E-01	2.6073701E 01	4.8466299E-01	-1.2230045E 00
0	43	2.9318019E-01	2.5993510E 01	4.8611175E-01	5.8254132E-01
0	44	8.3273214E-01	2.5300067E 01	4.8177869E-01	1.7284537E 00
0	45	2.4392665E-01	2.5240567E 01	4.8343453E-01	5.0457017E-01
0	46	3.4209192E-01	2.5123540E 01	4.8456107E-01	7.0598308E-01
0	47	-1.1776633E 00	2.3736649E 01	4.7321311E-01	-2.4886532E 00
0	48	-6.9730944E-01	2.3249571E 01	4.7055763E-01	-1.4831540E 00
0	49	4.8534400E-01	2.3014013E 01	4.7041320E-01	1.0317398E 00
0	50	1.8633364E-01	2.2979292E 01	4.7233454E-01	3.9449505E-01
0	51	-2.8886722E-01	2.2895848E 01	4.7378169E-01	-6.0970533E-01
0	52	1.5939509E 00	2.0355168E 01	4.4892796E-01	3.5505716E 00
1	0	1.1545718E 01	1.2457914E 02	9.0235400E-01	1.2795110E 01

75

Hourly Values of Ionospheric Characteristics at Screen Points

LAMBDA= 90.0 THETA= 0. (Geographic Coordinates of Screen Point)

6.13	6.08	6.08	6.33	6.36	6.13	6.00	6.00	5.87	5.77	5.77	5.76
5.68	5.55	5.53	5.72	5.98	6.23	6.34	6.23	6.16	6.25	6.22	6.12

LAMBDA= 80.0 THETA= 337.0

5.71	5.63	5.62	5.79	5.81	5.58	5.44	5.43	5.38	5.35	5.39	5.42
5.42	5.40	5.48	5.70	5.93	6.10	6.15	6.03	5.95	5.99	5.90	5.75

LAMBDA= 65.0 THETA= 157.0

6.03	5.87	5.76	5.72	5.80	5.46	6.04	6.08	6.21	6.35	6.42	6.49
6.51	6.49	6.53	6.57	6.57	6.62	6.68	6.64	6.59	6.53	6.39	6.21

LAMBDA= 40.0 THETA= 210.0

6.97	6.72	6.48	6.18	6.07	6.44	7.07	7.55	7.79	8.01	8.24	8.36
8.35	8.27	8.07	7.75	7.43	7.30	7.36	7.47	7.45	7.37	7.32	7.21

LAMBDA= 40.0 THETA= 330.0

7.51	7.05	6.57	6.13	5.95	6.33	7.04	7.64	8.01	8.31	8.60	8.76
8.80	8.79	8.71	8.51	8.31	8.26	8.38	8.51	8.48	8.35	8.21	7.94

LAMBDA= 35.0 THETA= 30.0

9.62	9.23	8.66	8.03	7.62	7.78	8.49	9.08	9.25	9.43	9.84	10.27
10.64	10.97	11.12	11.09	10.96	10.80	10.61	10.32	9.92	9.68	9.72	9.77

LAMBDA= 35.0 THETA= 105.0

9.88	9.64	9.16	8.57	8.16	8.29	8.91	9.38	9.41	9.49	9.89	10.38
10.84	11.24	11.43	11.39	11.23	10.97	10.64	10.18	9.66	9.41	9.57	9.82

LAMBDA= 35.0 THETA= 185.0

8.34	8.23	7.39	7.64	7.37	7.52	8.12	8.60	8.68	8.78	9.12	9.43
9.64	9.76	9.68	9.41	9.10	8.85	8.70	8.53	8.27	8.13	8.22	8.34

LAMBDA= 35.0 THETA= 310.0

7.96	7.62	7.26	6.86	6.58	6.78	7.50	8.17	8.45	8.70	9.05	9.27
9.35	9.37	9.24	8.97	8.72	8.59	8.66	8.73	8.65	8.52	8.44	8.26

LAMBDA= 30.0 THETA= 165.0

9.53	9.34	8.92	8.40	7.86	7.73	8.31	8.88	8.89	9.00	9.57	10.20
10.75	11.19	11.33	11.25	11.07	10.82	10.53	10.09	9.54	9.26	9.33	9.49

76

Second Analysis

s	k	d_k ORTHO COEFF.	E_k SUMS EK	e_k RMS (EK)	d_k/e_k ORTHO COEFF/RMS
0	0	1.1586693E 02	9.1748523E 02	2.0803270E 00	5.5696498E 01
0	1	2.7317004E 00	9.1002303E 02	2.0767536E 00	1.3153705E 00
0	2	-2.7691270E 01	1.4321663E 02	8.2582319E-01	-3.3531717E 01
0	3	-1.0710768E 00	1.4206942E 02	8.2447437E-01	-1.2991025E 00
0	4	1.8120750E 00	1.3878580E 02	8.1684725E-01	2.2183768E 00
0	5	1.3887605E 00	1.3685715E 02	8.1310862E-01	1.7079643E 00
0	6	2.1874004E-02	1.3685667E 02	8.1507837E-01	2.6836688E-02
0	7	-9.0152808E-01	1.3604391E 02	8.1463417E-01	-1.1066662E 00
0	8	4.1653732E-01	1.3587041E 02	8.1610746E-01	5.1039518E-01
0	9	-1.6968181E-01	1.3584162E 02	8.1802843E-01	-2.0742777E-01
0	10	-2.5998666E-01	1.3577402E 02	8.1984670E-01	-3.1711619E-01
0	11	1.2061595E 00	1.3431920E 02	8.1746848E-01	1.4754813E 00
0	12	1.7126742E 00	1.3138629E 02	8.1051307E-01	2.1129507E 00
0	13	-4.7527678E-01	1.3116040E 02	8.1184818E-01	-5.8542568E-01
0	14	3.4918428E 00	1.1896743E 02	7.7514232E-01	4.5047764E 00
0	15	2.8463537E 00	1.1086570E 02	7.5018002E-01	3.7942275E 00
0	16	5.8259827E 00	7.6923630E 01	6.2647226E-01	9.2996657E 00
0	17	-1.0570359E 00	7.5806304E 01	6.2349841E-01	-1.6953306E 00
0	18	5.7026108E-02	7.5803052E 01	6.2508989E-01	9.1228651E-02
0	19	2.1162752E-01	7.5758265E 01	6.2652204E-01	3.3778145E-01
0	20	-1.7289138E 00	7.2769122E 01	6.1563450E-01	-2.8083445E 00
0	21	-9.5828191E-01	7.1850817E 01	6.1333700E-01	-1.5624068E 00
0	22	1.2780304E 00	7.0217456E 01	6.0791903E-01	2.1023037E 00
0	23	2.9435686E-01	7.0130810E 01	6.0914898E-01	.4832263BE-01
0	24	-1.4432488E 00	6.8047842E 01	6.0162829E-01	-2.3989044E 00
0	25	-7.1608432E-01	6.7535064E 01	6.0095763E-01	-1.1915721E 00
0	26	-1.4521610E 00	6.5426322E 01	5.9308887E-01	-2.4484543E 00
0	27	2.4560481E 00	6.5366000E 01	5.9441544E-01	4.1318712E 00
0	28	2.0283353E 00	6.1251855E 01	5.7696665E-01	3.5155157E 00
0	29	8.5172019E-01	6.0526428E 01	5.7510477E-01	1.4809826E 00
0	30	1.1125955E 00	5.9288559E 01	5.7075503E-01	1.9493397E 00
0	31	2.0158679E-01	5.9247922E 01	5.7213335E-01	3.5234231E-01
0	32	-2.1212865E 00	5.4748065E 01	5.5150332E-01	-3.8463712E 00
0	33	-9.7172962E-01	5.3803806E 01	5.4825169E-01	-1.7724152E 00
0	34	-9.4321649E-01	5.2914149E 01	5.4522518E-01	-1.7299577E 00
0	35	1.3584320E 00	5.1068811E 01	5.3714463E-01	2.5289873E 00
0	36	1.4552336E 00	4.8951106E 01	5.2738153E-01	2.7593565E 00
0	37	7.0770034E-01	4.8450266E 01	5.2617360E-01	1.344994OE 00
0	38	2.7800587E-01	4.8372979E 01	5.2726238E-01	5.2726285E-01
0	39	9.6084980E-01	4.7449746E 01	5.2371365E-01	1.8346854E 00
0	40	-9.1351080E-01	4.6615244E 01	5.2059472E-01	-1.7547447E 00
0	41	-3.0404171E-01	4.6522802E 01	5.2159674E-01	-5.8290569E-01
0	42	-8.6670141E-01	4.5771630E 01	5.1888812E-01	-1.6703050E 00
0	43	1.9961724E-01	4.5731783E 01	5.2019445E-01	3.8373581E-01
0	44	1.2900269E-01	4.5715141E 01	5.2164540E-01	2.4729959E-01
0	45	6.7134166E-01	4.5264441E 01	5.2061940E-01	1.2895056E 00
0	46	1.2942037E 00	4.3589478E 01	5.1243264E-01	2.5256074E 00
0	47	-1.3200679E 00	4.1846898E 01	5.0360457E-01	-2.6212389E 00
0	48	-1.0531998E 00	4.0737668E 01	4.9839785E-01	-2.1131709E 00
0	49	-8.8752718E-01	3.9949963E 01	4.9506747E-01	-1.7927399E 00
0	50	-1.6098626E-01	3.9924047E 01	4.9643200E-01	-3.2428662E-01
0	51	1.0323948E 00	3.8858207E 01	4.9127927E-01	2.1014417E 00
0	52	1.5491854E 00	3.6458232E 01	4.7735097E-01	3.2453802E 00
1	0	1.5186295E 01	1.6917869E 02	8.9331555E-01	1.6999922E 01
1	1	-5.2349702E 00	1.4177378E 02	8.1970338E-01	-6.3864201E 00
1	2	-8.2212824E 00	7.4184297E 01	5.9435557E-01	-1.3832262E 01
1	3	2.3026811E 00	6.8881957E 01	5.7408948E-01	4.0110143E 00
1	4	-3.1106774E-01	6.8785193E 01	5.7506351E-01	-5.4092762E-01
1	5	1.9968812E 00	6.4797658E 01	5.5949277E-01	3.5690922E 00

77

Residuals and their RMS

00687   A DATA
LAMBDA= 86.8   THETA= 219.9 (Geographic coordinates)

H LZT	GAMMA	RESIDUAL	Y
0	6.13	-0.23	5.90
1	6.06	-0.06	6.00
2	6.13	-0.13	6.00
3	6.22	-0.32	5.90
4	5.96	-0.06	5.90
5	5.64	-0.14	5.50
6	5.63	0.07	5.70
7	5.63	-0.03	5.60
8	5.51	-0.01	5.50
9	5.56	-0.06	5.50
10	5.68	-0.08	5.60
11	5.70	-0.20	5.50
12	5.67	-0.27	5.40
13	5.58	-0.28	5.30
14	5.51	-0.51	5.00
15	5.61	-0.61	5.00
16	5.77	-0.27	5.50
17	5.86	-0.86	5.00
18	5.85	-0.75	5.10
19	5.81	-0.81	5.00
20	5.92	-0.62	5.30
21	6.10	-0.50	5.60
22	6.14	-0.54	5.60
23	6.14	-0.24	5.90

LZT = Zone time

Y = Observed value

Gamma = Computed value

Residual = Y − Gamma

LOCAL RMS RESIDUAL
4.105988E-01

00483   A DATA
LAMBDA= 82.3   THETA= 141.7

H LZT	GAMMA	RESIDUAL	Y
0	5.37	-0.17	5.20
1	5.39	-0.19	5.20
2	5.27	-0.17	5.10
3	5.29	-0.09	5.20
4	5.49	0.01	5.50
5	5.63	-0.03	5.60
6	5.67	0.33	6.00
7	5.57	0.23	5.80
8	5.37	0.63	6.00
9	5.30	0.00	5.30
10	5.39	-0.39	5.00
11	5.54	-0.04	5.50
12	5.62	-0.02	5.60
13	5.63	0.17	5.80
14	5.68	0.12	5.80
15	5.79	0.21	6.00
16	5.81	0.09	5.90
17	5.80	-0.10	5.70
18	5.81	-0.01	5.80
19	5.75	0.05	5.80
20	5.75	-0.05	5.70
21	5.85	-0.15	5.70
22	5.71	0.09	5.80
23	5.41	0.39	5.80

LOCAL RMS RESIDUAL
2.144553E-01

```
00982      A DATA
LAMBDA= 79.9   THETA= 243.9
```

H LZT	GAMMA	RESIDUAL	Y
0	5.61	0.09	5.70
1	5.66	-0.06	5.60
2	5.65	0.15	5.80
3	5.67	0.18	5.85
4	5.63	0.22	5.85
5	5.61	-0.01	5.60
6	5.71	-0.21	5.50
7	5.67	-0.17	5.50
8	5.45	0.15	5.60
9	5.43	-0.03	5.40
10	5.56	-0.06	5.50
11	5.53	-0.03	5.50
12	5.47	0.13	5.60
13	5.50	-0.00	5.50
14	5.49	0.21	5.70
15	5.45	0.10	5.55
16	5.41	0.14	5.55
17	5.36	0.04	5.40
18	5.40	0.30	5.70
19	5.54	0.11	5.65
2C	5.64	-0.24	5.40
21	5.64	0.06	5.70
22	5.55	0.05	5.60
23	5.52	-0.02	5.50

```
LOCAL RMS RESIDUAL
  1.416085E-01
```

```
0009-      A DATA
LAMBDA=-90.0   THETA=   0.
```

H LZT	GAMMA	RESIDUAL	Y
0	4.68	0.12	4.80
1	4.80	0.40	5.20
2	5.08	0.02	5.10
3	5.32	-0.12	5.20
4	5.23	0.77	6.00
5	5.03	0.37	5.40
6	4.96	0.14	5.10
7	4.70	0.50	5.20
8	4.45	0.85	5.30
9	4.77	0.53	5.30
10	5.23	-0.03	5.20
11	5.23	-0.03	5.20
12	5.08	-0.08	5.00
13	5.03	-0.13	4.90
14	4.95	0.25	5.20
15	4.88	0.02	4.90
16	4.61	0.39	5.00
17	3.94	1.56	5.50
18	3.35	1.35	4.70
19	3.29	1.11	4.40
2C	3.58	0.92	4.50
21	4.00	0.70	4.70
22	4.39	0.41	4.80
23	4.60	0.40	5.00

```
LOCAL RMS RESIDUAL
  6.320838E-01

RMS RESIDUAL     A DATA     79
  6.318335E-01
```

```
00471      B DATA
LAMBDA= 71.6   THETA= 128.9
H LZT      GAMMA        RESIDUAL       Y
    C       5.77         0.63         6.40
    1       5.73         0.67         6.40
    2       5.66         0.74         6.40
    3       5.66         0.64         6.30
    4       5.72         0.38         6.10
    5       5.68         0.22         5.90
    6       5.64         0.16         5.80
    7       5.68         0.12         5.80
    8       5.72         0.08         5.80
    9       5.76         0.04         5.80
   10       5.90         0.00         5.90
   11       6.08        -0.08         6.00
   12       6.21        -0.21         6.00
   13       6.26        -0.16         6.10
   14       6.20         0.00         6.20
   15       6.06         0.34         6.40
   16       5.96         0.54         6.50
   17       5.93         0.67         6.60
   18       5.90         0.80         6.70
   19       5.84         0.86         6.70
   20       5.80         0.80         6.60
   21       5.82         0.68         6.50
   22       5.83         0.67         6.50
   23       5.80         0.60         6.40
LOCAL RMS RESIDUAL
  5.112251E-01
```

```
0047P      B DATA
LAMBDA=-78.4   THETA= 106.9
H LZT      GAMMA        RESIDUAL       Y
    0       4.25        -0.85         3.40
    1       3.90        -0.60         3.30
    2       3.59        -0.49         3.10
    3       3.40        -0.30         3.10
    4       3.33        -0.13         3.20
    5       3.48         0.02         3.50
    6       3.76         0.04         3.80
    7       3.94         0.26         4.20
    8       4.19         0.41         4.60
    9       4.88         0.22         5.10
   10       5.73        -0.13         5.60
   11       6.25        -0.05         6.20
   12       6.49         0.11         6.60
   13       6.55         0.15         6.70
   14       6.43         0.17         6.60
   15       6.37        -0.07         6.30
   16       6.33        -0.53         5.80
   17       6.06        -0.56         5.50
   18       5.65        -0.55         5.10
   19       5.34        -0.44         4.90
   20       5.05        -0.55         4.50
   21       4.77        -0.57         4.20
   22       4.62        -0.82         3.80
   23       4.50        -0.90         3.60

LOCAL RMS RESIDUAL
  4.569347E-01

RMS RESIDUAL      B DATA
  6.237257E-01

RMS RESIDUAL OF A AND B DATA
  6.312054E-01
```

One Per Card Coefficients D_{sk} (See Card Format 5)

Y M	DLC	N	P	H	FNCD	PP1	PP2	T	SN	s	k	D_{sk}
5806	4500	213	52	8	100	12	36	1	186.8	0	0	1.0847341E 01
5806	4500	213	52	8	100	12	36	1	186.8	0	1	6.0068659E-01
5806	4500	213	52	8	100	12	36	1	186.8	0	2	-2.4230472E 01
5806	4500	213	52	8	100	12	36	1	186.8	0	3	1.7365042E 01
5806	4500	213	52	8	100	12	36	1	186.8	0	4	1.8869887E 02
5806	4500	213	52	8	100	12	36	1	186.8	0	5	-1.1593593E 02
5806	4500	213	52	8	100	12	36	1	186.8	0	6	-7.6300942E 02
5806	4500	213	52	8	100	12	36	1	186.8	0	7	2.6676274E 02
5806	4500	213	52	8	100	12	36	1	186.8	0	8	1.4264726E 03
5806	4500	213	52	8	100	12	36	1	186.8	0	9	-2.5996887E 02
5806	4500	213	52	8	100	12	36	1	186.8	0	10	-1.2412525E 03
5806	4500	213	52	8	100	12	36	1	186.8	0	11	9.1843662E 01
5806	4500	213	52	8	100	12	36	1	186.8	0	12	4.0777261E 02
5806	4500	213	52	8	100	12	36	1	186.8	0	13	1.4352705E-01
5806	4500	213	52	8	100	12	36	1	186.8	0	14	9.6570256E-01
5806	4500	213	52	8	100	12	36	1	186.8	0	15	-8.4557281E-01
5806	4500	213	52	8	100	12	36	1	186.8	0	16	-4.7783220E 00
5806	4500	213	52	8	100	12	36	1	186.8	0	17	-1.3976016E 01
5806	4500	213	52	8	100	12	36	1	186.8	0	18	-2.3884222E 01
5806	4500	213	52	8	100	12	36	1	186.8	0	19	4.4348199E 01
5806	4500	213	52	8	100	12	36	1	186.8	0	20	1.1777619E 02
5806	4500	213	52	8	100	12	36	1	186.8	0	21	1.1766320E 02
5806	4500	213	52	8	100	12	36	1	186.8	0	22	1.5439476E 02
5806	4500	213	52	8	100	12	36	1	186.8	0	23	-2.9505143E 02
5806	4500	213	52	8	100	12	36	1	186.8	0	24	-5.8452889E 02
5806	4500	213	52	8	100	12	36	1	186.8	0	25	-3.3965383E 02
5806	4500	213	52	8	100	12	36	1	186.8	0	26	-3.8447369E 02
5806	4500	213	52	8	100	12	36	1	186.8	0	27	7.7576793E 02
5806	4500	213	52	8	100	12	36	1	186.8	0	28	1.2218275E 03
5806	4500	213	52	8	100	12	36	1	186.8	0	29	3.9903615E 02
5806	4500	213	52	8	100	12	36	1	186.8	0	30	4.1502135E 02
5806	4500	213	52	8	100	12	36	1	186.8	0	31	-8.8033071E 02
5806	4500	213	52	8	100	12	36	1	186.8	0	32	-1.1519065E 03
5806	4500	213	52	8	100	12	36	1	186.8	0	33	-1.6575612E 02
5806	4500	213	52	8	100	12	36	1	186.8	0	34	-1.6211853E 02
5806	4500	213	52	8	100	12	36	1	186.8	0	35	3.6065598E 02
5806	4500	213	52	8	100	12	36	1	186.8	0	36	4.0243153E 02
5806	4500	213	52	8	100	12	36	1	186.8	0	37	2.8422108E-01
5806	4500	213	52	8	100	12	36	1	186.8	0	38	3.9405910E-01
5806	4500	213	52	8	100	12	36	1	186.8	0	39	-1.9527271E 00
5806	4500	213	52	8	100	12	36	1	186.8	0	40	-3.6131784E 00
5806	4500	213	52	8	100	12	36	1	186.8	0	41	-5.2786767E 00
5806	4500	213	52	8	100	12	36	1	186.8	0	42	-4.8036404E 00
5806	4500	213	52	8	100	12	36	1	186.8	0	43	2.1448102E 01
5806	4500	213	52	8	100	12	36	1	186.8	0	44	3.3071747E 01
5806	4500	213	52	8	100	12	36	1	186.8	0	45	1.8113526E 01
5806	4500	213	52	8	100	12	36	1	186.8	0	46	9.1406965E 00
5806	4500	213	52	8	100	12	36	1	186.8	0	47	-4.7457672E 01
5806	4500	213	52	8	100	12	36	1	186.8	0	48	-8.1269085E 01
5806	4500	213	52	8	100	12	36	1	186.8	0	49	-1.5238267E 01
5806	4500	213	52	8	100	12	36	1	186.8	0	50	-3.3477439E 00
5806	4500	213	52	8	100	12	36	1	186.8	0	51	2.9248889E 01
5806	4500	213	52	8	100	12	36	1	186.8	0	52	5.5277724E 01
5806	4500	213	52	8	100	12	36	1	186.8	1	0	1.5654446E 00
5806	4500	213	52	8	100	12	36	1	186.8	1	1	-1.7650886E 00

One Per Card Coefficients h_{sk} (See Card Format 5)

Y M	DLC	N	P	H	FNCD	PP1	PP2	T	SN	s	k	h_{sk}
5806	4500	213	52	8	100	12	36	2	186.8	0	0	9.4565125E 01
5806	4500	213	52	8	100	12	36	2	186.8	0	1	4.0170246E 00
5806	4500	213	52	8	100	12	36	2	186.8	0	2	-2.1605705E 01
5806	4500	213	52	8	100	12	36	2	186.8	0	3	-7.8164387E-01
5806	4500	213	52	8	100	12	36	2	186.8	0	4	1.4642438E 00
5806	4500	213	52	8	100	12	36	2	186.8	0	5	1.3993547E 00
5806	4500	213	52	8	100	12	36	2	186.8	0	6	-3.2498074E-02
5806	4500	213	52	8	100	12	36	2	186.8	0	7	-3.7577779E-01
5806	4500	213	52	8	100	12	36	2	186.8	0	8	1.3522217E-01
5806	4500	213	52	8	100	12	36	2	186.8	0	9	-3.9923328E-01
5806	4500	213	52	8	100	12	36	2	186.8	0	10	-6.4115469E-01
5806	4500	213	52	8	100	12	36	2	186.8	0	11	7.1357547E-01
5806	4500	213	52	8	100	12	36	2	186.8	0	12	1.5940220E 00
5806	4500	213	52	8	100	12	36	2	186.8	0	13	-9.0769068E-01
5806	4500	213	52	8	100	12	36	2	186.8	0	14	1.8921489E 00
5806	4500	213	52	8	100	12	36	2	186.8	0	15	2.6680483E 00
5806	4500	213	52	8	100	12	36	2	186.8	0	16	5.1834916E 00
5806	4500	213	52	8	100	12	36	2	186.8	0	17	-6.0173756E-01
5806	4500	213	52	8	100	12	36	2	186.8	0	18	2.8737972E-01
5806	4500	213	52	8	100	12	36	2	186.8	0	19	7.5293761E-02
5806	4500	213	52	8	100	12	36	2	186.8	0	20	-1.5618081E 00
5806	4500	213	52	8	100	12	36	2	186.8	0	21	-8.1786337E-01
5806	4500	213	52	8	100	12	36	2	186.8	0	22	1.0628657E 00
5806	4500	213	52	8	100	12	36	2	186.8	0	23	3.9685505E-01
5806	4500	213	52	8	100	12	36	2	186.8	0	24	-7.7826549E-01
5806	4500	213	52	8	100	12	36	2	186.8	0	25	-3.5562921E-01
5806	4500	213	52	8	100	12	36	2	186.8	0	26	-1.2530024E 00
5806	4500	213	52	8	100	12	36	2	186.8	0	27	1.6143535E-01
5806	4500	213	52	8	100	12	36	2	186.8	0	28	1.2648626E 00
5806	4500	213	52	8	100	12	36	2	186.8	0	29	7.4794657E-01
5806	4500	213	52	8	100	12	36	2	186.8	0	30	1.2934358E 00
5806	4500	213	52	8	100	12	36	2	186.8	0	31	6.1983575E-02
5806	4500	213	52	8	100	12	36	2	186.8	0	32	-1.8562712E 00
5806	4500	213	52	8	100	12	36	2	186.8	0	33	-9.3592011E-01
5806	4500	213	52	8	100	12	36	2	186.8	0	34	-9.1538558E-01
5806	4500	213	52	8	100	12	36	2	186.8	0	35	1.0113917E 00
5806	4500	213	52	8	100	12	36	2	186.8	0	36	1.1285386E 00
5806	4500	213	52	8	100	12	36	2	186.8	0	37	3.4079790E-01
5806	4500	213	52	8	100	12	36	2	186.8	0	38	3.2231821E-01
5806	4500	213	52	8	100	12	36	2	186.8	0	39	8.8341281E-01
5806	4500	213	52	8	100	12	36	2	186.8	0	40	-9.1879728E-01
5806	4500	213	52	8	100	12	36	2	186.8	0	41	-9.5740420E-02
5806	4500	213	52	8	100	12	36	2	186.8	0	42	-6.3193738E-01
5806	4500	213	52	8	100	12	36	2	186.8	0	43	1.1002588E-01
5806	4500	213	52	8	100	12	36	2	186.8	0	44	-9.0577019E-02
5806	4500	213	52	8	100	12	36	2	186.8	0	45	5.0261891E-01
5806	4500	213	52	8	100	12	36	2	186.8	0	46	1.2591517E 00
5806	4500	213	52	8	100	12	36	2	186.8	0	47	-1.0611030E 00
5806	4500	213	52	8	100	12	36	2	186.8	0	48	-1.1026031E 00
5806	4500	213	52	8	100	12	36	2	186.8	0	49	-8.0437840E-01
5806	4500	213	52	8	100	12	36	2	186.8	0	50	-1.7631563E-01
5806	4500	213	52	8	100	12	36	2	186.8	0	51	8.2920978E-01
5806	4500	213	52	8	100	12	36	2	186.8	0	52	1.4233196E 00
5806	4500	213	52	8	100	12	36	2	186.8	1	0	1.4133058E 01
5806	4500	213	52	8	100	12	36	2	186.8	1	1	-3.7146698E 00

GPO 835-527

U. S. DEPARTMENT OF COMMERCE
Luther H. Hodges, *Secretary*

NATIONAL BUREAU OF STANDARDS
A. V. Astin, *Director*

THE NATIONAL BUREAU OF STANDARDS

The scope of activities of the National Bureau of Standards at its major laboratories in Washington, D.C., and Boulder, Colorado, is suggested in the following listing of the divisions and sections engaged in technical work. In general, each section carries out specialized research, development, and engineering in the field indicated by its title. A brief description of the activities, and of the resultant publications, appears on the inside of the front cover.

WASHINGTON, D. C.

Electricity. Resistance and Reactance. Electrochemistry. Electrical Instruments. Magnetic Measurements. Dielectrics. High Voltage.

Metrology. Photometry and Colorimetry. Refractometry. Photographic Research. Length. Engineering Metrology. Mass and Scale. Volumetry and Densimetry.

Heat. Temperature Physics. Heat Measurements. Cryogenic Physics. Equation of State. Statistical Physics.

Radiation Physics. X-ray. Radioactivity. Radiation Theory. High Energy Radiation. Radiological Equipment. Nucleonic Instrumentation. Neutron Physics.

Analytical and Inorganic Chemistry. Pure Substances. Spectrochemistry. Solution Chemistry. Standard Reference Materials. Applied Analytical Research. Crystal Chemistry.

Mechanics. Sound. Pressure and Vacuum. Fluid Mechanics. Engineering Mechanics. Rheology. Combustion Controls.

Polymers. Macromolecules: Synthesis and Structure. Polymer Chemistry. Polymer Physics. Polymer Characterization. Polymer Evaluation and Testing. Applied Polymer Standards and Research. Dental Research.

Metallurgy. Engineering Metallurgy. Microscopy and Diffraction. Metal Reactions. Metal Physics. Electrolysis and Metal Deposition.

Inorganic Solids. Engineering Ceramics. Glass. Solid State Chemistry. Crystal Growth. Physical Properties. Crystallography.

Building Research. Structural Engineering. Fire Research. Mechanical Systems. Organic Building Materials. Codes and Safety Standards. Heat Transfer. Inorganic Building Materials. Metallic Building Materials.

Applied Mathematics. Numerical Analysis. Computation. Statistical Engineering. Mathematical Physics. Operations Research.

Data Processing Systems. Components and Techniques. Computer Technology. Measurements Automation. Engineering Applications. Systems Analysis.

Atomic Physics. Spectroscopy. Infrared Spectroscopy. Far Ultraviolet Physics. Solid State Physics. Electron Physics. Atomic Physics. Plasma Spectroscopy.

Instrumentation. Engineering Electronics. Electron Devices. Electronic Instrumentation. Mechanical Instruments. Basic Instrumentation.

Physical Chemistry. Thermochemistry. Surface Chemistry. Organic Chemistry. Molecular Spectroscopy. Elementary Processes. Mass Spectrometry. Photochemistry and Radiation Chemistry.

Office of Weights and Measures.

BOULDER, COLO.

Cryogenic Engineering Laboratory. Cryogenic Equipment. Cryogenic Processes. Properties of Materials. Cryogenic Technical Services.

CENTRAL RADIO PROPAGATION LABORATORY

Ionosphere Research and Propagation. Low Frequency and Very Low Frequency Research. Ionosphere Research. Prediction Services. Sun-Earth Relationships. Field Engineering. Radio Warning Services. Vertical Soundings Research.

Radio Propagation Engineering. Data Reduction Instrumentation. Radio Noise. Tropospheric Measurements. Tropospheric Analysis. Propagation-Terrain Effects. Radio-Meteorology. Lower Atmosphere Physics.

Radio Systems. Applied Electromagnetic Theory. High Frequency and Very High Frequency Research. Frequency Utilization. Modulation Research. Antenna Research. Radiodetermination.

Upper Atmosphere and Space Physics. Upper Atmosphere and Plasma Physics. High Latitude Ionosphere Physics. Ionosphere and Exosphere Scatter. Airglow and Aurora. Ionospheric Radio Astronomy.

RADIO STANDARDS LABORATORY

Radio Physics. Radio Broadcast Service. Radio and Microwave Materials. Atomic Frequency and Time-Interval Standards. Radio Plasma. Millimeter-Wave Research.

Circuit Standards. High Frequency Electrical Standards. High Frequency Calibration Services. High Frequency Impedance Standards. Microwave Calibration Services. Microwave Circuit Standards. Low Frequency Calibration Services.

www.ingramcontent.com/pod-product-compliance
Lightning Source LLC
Chambersburg PA
CBHW061016050326
40689CB00012B/2665

* 9 7 8 1 0 1 4 7 5 8 6 8 2 *